Ultrasound Contrast Agents

Gaio Paradossi, Paolo Pellegretti, Andrea Trucco (Eds.)

Ultrasound Contrast Agents

Targeting and Processing Methods
for Theranostics

 Springer

Editors

Gaio Paradossi

Dipartimento di Scienze e Tecnologie Chimiche
Università di Roma Tor Vergata
Rome, Italy

Paolo Pellegretti

Esaote – R & D Ultrasuoni
Genoa, Italy

Andrea Trucco

Acoustics, Antenna Arrays, and Underwater Signals Lab.
Dept. of Biophysical and Electronic Engineering (DIBE)
University of Genoa
Genoa, Italy

ISBN 978-88-470-1493-0   e-ISBN 978-88-470-1494-7
DOI 10.1007/978-88-470-1494-7

Springer Dordrecht Heidelberg London Milan New York

Library of Congress Control Number: 2009938492

9 8 7 6 5 4 3 2 1

Cover-Design: Somona Colombo (Milan)
Typesetting: PTP-Berlin, Protago TeX-Production GmbH, Germany (www.ptp-berlin.eu)
Printing and Binding: Grafiche Porpora, Segrate (Milan)
Printed in Italy

Springer-Verlag Italia srl – Via Decembrio 28 –20137 Milan
Springer is a part of Springer Science+Business Media (www.springer.com)

Preface

Medical ultrasound is one of the most used diagnostic tools in medicine. This well assessed imaging technique has been the object of very intense research activity over the last two decades, and this scientific burst of interest does not show any signs of decrease. This interest stems essentially from two main reasons:

(i) Medical ultrasound is a handy and relatively low-cost methodology and is spread over a number of health care fields.

(ii) The development of new technologies for the fabrication of micro/nano devices implemented by remarkable achievements in hardware and software capabilities has uncovered new potentialities for already established medical methodologies such as medical ultrasound.

As a consequence, in recent years the production of scientific papers and patents concerning different aspects of the research in this field has maintained very high standards outlining an ever-changing scientific scenario where the concept of ultrasound contrast agents has progressively moved from a purely diagnostic function to a multifunctional device supporting also a therapeutic approach. This combined strategy is only part of a broader medical approach to most common pathologies where Europe is fully engaged in this highly competitive arena, with the participation of scientists and technologists from academia and industry.

We hope this edited book will give an updated picture of the main achievements accomplished in the field of next-generation multifunctional ultrasound contrast agents within the framework of two projects (TAMIRUT and SIGHT) funded by the European Commission in the context of the sixth Framework Programme. The authors of the chapters attempted to address all steps of a new integrated strategy, from the development of new echogenic injectable devices, to the interactions with cells and tissues, including the development of *ad hoc* hardware and signal processing techniques.

The book is organized as follows. The first chapter is an introduction, describing the possible contribution of ultrasound contrast agents (in their present and future forms) towards the application of molecular imaging in clinical practice, and the convergence between diagnosis and therapy.

Chapters 2 to 4 address the design of next-generation ultrasound contrast agents. Methods to support new features such as targeting and *in situ* drug delivery are described in more detail for lipidic and polymer shelled microparticles. Injectable devices require an assessment of their biocompatibility and, more generally, of the study of the biointerface which occurs when cells or tissues are exposed to the synthetic microdevices. In chapters 5 and 6, issues such as the protein coating of microballoon surfaces when in contact with biological fluids are discussed, and the study of the biocompatibility and internalization processes in fibroblasts and macrophages are presented.

Chapters 7 to 9 focus on some new aspects concerning the modeling and characterization of ultrasound contrast agents. In particular, chapter 7 investigates (theoretically and experimentally) the influence of a neighboring wall on the dynamics of coated microbubbles. This is of particular interest for predicting the acoustic behavior of target microbubbles bound to pathological cells. Chapter 8 examines the acoustic behavior and disruption of innovative PVA-shelled microballoons (potentially, a new generation of ultrasound contrast agents), as a function of temperature. The same kind of microdevice is considered in chapter 9, where its physical characterization through atomic force microscopy (AFM), reflection interference contrast microscopy, and scanning transmission soft X-ray microspectroscopy is presented.

Chapters 10 to 13 introduce some recent achievements in medical ultrasound equipment and signal processing, concerning an enhanced exploitation of contrast agents. Specifically, chapter 10 aims to highlight the difference in the performance obtained by multi-pulse techniques when realistic undesired effects are taken into account, and to present a simulation tool which allows this investigation to be carried out. Chapter 11 has a pioneering goal: to measure the volumetric concentration of the contrast agent inside the body, acting exclusively on the signals remotely acquired during a medical ultrasound examination. This could provide, when coupled with targeted microbubbles, new diagnostic information of great importance. Chapter 12 presents the Bubble Behavior Testing system, an ideal electronic tool to experimentally investigate the behavior of microbubbles excited by an ultrasound field. Possible applications are illustrated, also in combination with a high-speed optical camera. Finally, chapter 13 depicts the electronic architecture of a modern high-end medical ultrasound scanner, designed to allow easy experimentation with present and future techniques aimed at improving contrast enhanced imaging.

G. Paradossi
P. Pellegretti
A. Trucco

Acknowledgments

This edited book is a direct consequence of an international workshop organized as a cross-fertilization and dissemination action in the context of two research projects funded by the European Commission:

- SIGHT – Systems for in-situ theranostics using micro-particles triggered by ultrasound, FP6-IST IST-2005-2.5.2, 2006-20010.
- TAMIRUT – A new bio-sensor concept for medical diagnosis: targeted micro-bubbles and remote ultrasound transduction, FP6-NMP4-CT-2005-016382, 2005-2008.

On behalf of all the authors, the Editors would like to acknowledge the support of the European Commission (through the funds for dissemination actions of the projects above mentioned) and Esaote S.p.A. (through direct financial sponsorship), which makes this publication possible, and they would like to express their gratitude for that.

Contents

List of Contributors

Allémann, Eric
School of Pharmaceutical Sciences
University of Geneva
University of Lausanne
Switzerland
eric.allemann@unige.ch

Bettinger, Thierry
Bracco Research SA
Geneva, Switzerland
thierry.bettinger@brg.bracco.com

Brismar, Torkel
Department of Radiology
Karolinska Institutet
Stockholm, Sweden
torkel.brismar@karolinska.se

Cedervall, Tommy
Centre for BioNano Interactions
School of Chemistry and Chemical
Biology
University College Dublin
Dublin, Ireland
t.cedervall@gmail.com

Cerroni, Barbara
Dipartimento di Scienze e Tecnologie
Chimiche
Università di Roma Tor Vergata
Roma, Italy
b.cerroni@gmail.com

Crocco, Marco
Dept. of Biophysical and Electronic
Engineering
University of Genoa, Italy
crocco@dibe.unige.it

Dawson, Kenneth
Centre for BioNano Interactions
School of Chemistry and Chemical
Biology
University College Dublin
Dublin, Ireland
kenneth@fiachra.ucd.ie

de Jong, Nico
Biomedical Engineering
Thorax Centre, Erasmus MC
Rotterdam, The Netherlands
n.dejong@erasmusmc.nl

Fernandes, Paulo
Max-Planck-Institute for
Colloids and Interfaces (MPIKG)
Wissenschaftspark Golm
Potsdam, Germany
Paulo.Fernandes@mpikg.mpg.de

Fery, Andreas
Department of Physical Chemistry II
University of Bayreuth
Bayreuth, Germany
andreas.fery@uni-bayreuth.de

Fink, Rainer
Friedrich-Alexander Universität
Erlangen
Physikalische Chemie II
Erlangen, Germany
Rainer.Fink@chemie.uni-
erlangen.de

Gajovic-Eichelmann, Nenad
Fraunhofer Institute for Biomedical
Engineering
Potsdam-Golm, Germany
nenad.gajovic@ibmt.fhg.de

Grishenkov, Dmitry
Marcus Wallenberg Laboratory
Royal Institute of Technology
Stockholm, Sweden
dmitryg@kth.se

Guidi, Francesco
Electronic Engineering Department
University of Florence
Florence, Italy
Francesco.Guidi@unifi.it

Hauwel, Mathieu
Bracco Research SA
Geneva, Switzerland
mathieu.hauwel@brg.bracco.com

Kounoudes, Anastasis
SignalGeneriX Ltd, Limassol
Cyprus
a.kounoudes@signalgenerix.com

Lionetti, Vincenzo
Sector of Medicine
Scuola Superiore Sant'Anna
Pisa, Italy
v.lionetti@sssup.it

Lundqvist, Martin
Centre for BioNano Interactions
School of Chemistry and Chemical
Biology
University College Dublin
Dublin, Ireland

Lynch, Iseult
Centre for BioNano Interactions
School of Chemistry and Chemical
Biology
University College Dublin
Dublin, Ireland
marlusphere@gmail.com

Mori, Riccardo
Electronic Engineering Department
University of Florence
Florence, Italy
Riccardo.Mori@unifi.it

Margheritelli, Silvia
Dipartimento di Scienze e Tecnologie
Chimiche
Università di Roma Tor Vergata
Roma, Italy
silvia_marg@libero.it

Mozetic, Pamela
Dipartimento di Scienze e Tecnologie
Chimiche
Università di Roma Tor Vergata
Roma, Italy
pamela.mozetic@uniroma2.it

Nencioni, Alessandro
Esaote S.p.A.
Florence, Italy
Alessandro.Nencioni@esaote.com

Oddo, Letizia
Dipartimento di Scienze e Tecnologie
Chimiche
Università di Roma Tor Vergata
Roma, Italy
letiziaoddo@alice.it

Overvelde, Marlies
Physics of Fluids Group
University of Twente
The Netherlands
M.L.J.Overvelde@tnw.utwente.nl

Paddeu, Sergio
Esaote S.p.A.
Genoa, Italy
Sergio.Paddeu@esaote.com

Paradossi, Gaio
Dipartimento di Scienze e Tecnologie
Chimiche
Università di Roma Tor Vergata
Roma, Italy
paradossi@stc.uniroma2.it

Pecorari, Claudio
Flisavägen 60, 72353 Västerås
Sweden
claudio.pecorari@hotmail.com

Pellegretti, Paolo
Esaote S.p.A.
Genoa, Italy
Paolo.Pellegretti@esaote.com

Pretzl, Melanie
Department of Physical Chemistry II
University of Bayreuth
Bayreuth, Germany
melanie.pretzl@uni-bayreuth.de

Sciallero, Claudia
Dept. of Biophysical and Electronic
Engineering
University of Genoa
Genoa, Italy
sciallero@ginevra.dibe.unige.it

Stigler, Johannes
Centre for BioNano Interactions
School of Chemistry and Chemical
Biology
University College Dublin
Dublin, Ireland
johannes.stigler@gmail.com

Tortoli, Piero
Electronic Engineering Department
University of Florence
Florence, Italy
Piero.Tortoli@unifi.it

Tortora, Mariarosaria
Dipartimento di Scienze e Tecnologie
Chimiche
Università di Roma Tor Vergata
Roma, Italy
mariarosaria.tortora@uniroma2.it

Trucco, Andrea
Dept. of Biophysical and Electronic
Engineering
University of Genoa
Genoa, Italy
trucco@dibe.unige.it

Tzvetkov, George
Department of Inorganic Chemistry
Faculty of Chemistry
University of Sofia 1
Sofia, Bulgaria
George.Tzvetkov@wmail.chem.uni-
sofia.bg

Versluis, Michel
Physics of Fluids Group
University of Twente
The Netherlands
m.versluis@utwente.nl

Vos, Hendrik J.
Biomedical Engineering
Thorax Centre, Erasmus MC
Rotterdam, The Netherlands
h.j.vos@erasmusmc.nl

Zerbe, Ines
Fraunhofer Institute for Biomedical
Engineering
Potsdam-Golm, Germany
ines.zerbe@ibmt.fraunhofer.de

Chapter 1

Towards Ultrasound Molecular Imaging

Vincenzo Lionetti and Sergio Paddeu

Abstract. Health systems are facing both health promotion and disease prevention thus requiring the discovery of new sustainable clinical and technological approaches concerning both (early) diagnosis and (personalized) therapy. This chapter aims to highlight the main steps forward toward the ultrasound molecular imaging, a new emerging technological approach which will contribute to accomplish the convergence of diagnosis and therapy in clinical practice. This convergence, which is called "theranostics", is currently being developed as a combined diagnostic-therapeutic clinical platform whereas integrated therapy systems involve diagnostic technologies like ultrasound, magnetic resonance or computed tomography. Ultrasound molecular imaging represents a worthwhile step toward the full exploitation of imaging technologies in aiding treatment and surgery through the exploitation of combining molecular biology and emerging technologies, like nanotechnology. Although molecular imaging has been mainly concerned with nuclear imaging, magnetic resonance or optical imaging, today there is a growing interest in the exploitation of diagnostic ultrasound due to the possible merging of ultrasound advantages over other imaging modalities like real time, non-invasiveness, low cost, therapy and diagnosis application, short and efficient imaging protocol. The growing list of clinical conditions concerning the diagnostic potential of ultrasound include several areas like angiogenesis, atherosclerotic plaque investigation, inflammation marker detection, and identification, and other. Moreover, the progress of nanotechnology (nano-medicine) as well as of molecular biology supported the recent development and engineering of ultrasound contrast agents, based on gas filled microbubbles or nanoparticles providing new insights into early tumor detection (e.g., angiogenesis), local drug delivery, early responses to molecular therapies and in situ therapy. This approach is favoring the growth theranostics, namely the set-up of novel strategies about diagnosis, drug development and therapy, in fields like oncology, cardiology or rheumatology.

1.1 Introduction

The term *molecular imaging* has been broadly applied to describe a heterogeneous family of non-invasive imaging techniques that have been developed for

Paradossi, G., Pellegretti, P., Trucco, A. (Eds.)
Ultrasound contrast agents. Targeting and processing methods for theranostics
© Springer-Verlag Italia, 2010

use in both basic science and clinical settings. For the molecular or cell biologist, this term encompasses the myriad of different optical techniques that have been developed to profile cell properties such as protein expression, metabolic status, intracellular molecular trafficking, gene transcription, enzyme activity, and pH. Such a technology gained popularity due to its ability to image very fine molecular changes opening exciting opportunities for medical application, including not only the early detection but also the treatment of disease and pharmaceutical development as well. The techniques employed for these diverse molecular imaging applications differ in terms of the capabilities of the detectors (spatial resolution and sensitivity); properties of the imaging probes (toxicity, biodistribution and specificity for target); and practical issues such as image acquisition time and expense.

Thus, different modalities can be used for non-invasive molecular imaging in vivo and the key modalities for non-invasive molecular imaging are PET (Positron Emission Tomography), SPECT (Single Photon Emission Computed Tomography) and Magnetic Resonance Imaging (MRI). Targeted ultrasound modality is likely starting to play a role in these molecular applications due to resolution, low-cost and availability compared to SPECT, PET, MRI, Computed Tomography (CT) and Optical Imaging [21].

It is well known that diagnostic ultrasound (US), namely the application of ultrasound to medical diagnosis, grew out of industrial non-destructive testing. Such a growth is witnessed both in instrumentation and its technical sophistication as well as in the range of clinical application [39]. Moreover, the US development is likely to continue into the foreseeable future ensured by the high versatility of the tool as well as of related and advanced research areas, like nanotechnology. In fact, nanoparticles can be engineered as nanoplatforms for effective and targeted imaging labels, also by using diagnostic ultrasounds, and delivery of drugs by overcoming the many biological, biophysical, and biomedical barriers.

In general, the use of a molecularly targeted nanoplatform affords many advantages over conventional approaches. First, more imaging labels or a combination of labels for different imaging modalities can be attached to a single nanoparticle, which can lead to merge different signals from the same target tissue in vivo. Second, multiple, potentially different, targeting ligands on the nanoparticle can provide enhanced binding affinity and specificity. Third, the ability to integrate a means to bypass biological barriers can enable enhanced targeting efficacy. Ultimately, the combination of different targeting ligands, imaging labels, therapeutic drugs, and many other agents may allow for effective and controlled delivery of therapeutic agents in patients, which can be non-invasively monitored in real time.

The visualization of biological processes in living systems has the main expectation in imaging and quantification of molecular changes associated with a disease at early stage rather than the resulting morphological changes. It could also be used to guide treatment in patients and should facilitate the development and testing of new therapies in the laboratory setting.

Ultrasounds are almost never associated with molecular imaging, but the recent advances in ultrasound contrast agents (UCA) and related technologies have developed a well balanced diagnostic approach in terms of sensitivity and spatial resolution. Molecular imaging with US uses mainly intravascular delivery of microbubbles (UCA) that are targeted to bind to disease-specific epitopes expressed by endothelial cells. The binding of the UCA to the endothelium after systemic injection occurs by virtue of targeting ligands on the microparticle surface that confer specific binding to the endothelial target. In the presence of US, these bound bubbles produce an echo signal that can be detected as a persistent contrast effect on a 2-dimensional US image. Several strategies have been used to target UCA to regions of disease. The first strategy simply takes advantage of inherent chemical or electrostatic properties of the microbubble shell that promote retention of microbubbles within diseased organs [23,24]. A second and more selective strategy relies on the attachment of antibodies, peptides, or other ligands to the microbubble surface that recognize disease-related antigens [16]. Another ligand-directed strategy using non-bubble lipid emulsion UCA relies on administration of biotinylated monoclonal antibodies, followed by injection of streptavidin, to which biotinylated emulsion contrast agents attach [25].

For the specific purposes of imaging molecular phenotype, it is necessary that the imaging technique be sensitive, be selective for the molecular event targeted, and provides good spatial resolution. The possibility of molecular imaging via US is substantiated by a growing body of literature demonstrating proof of concept in vitro and in clinically relevant animal models of disease, lending support to the notion that the "bench-to-imaging" translation of this technology will be possible in the near future.

1.2 Ultrasound molecular probes: new features of imaging

The use of UCA for molecular imaging is an extension of contrast echographic principles and approaches that are already in clinical practice. All the US contrast agents have in common the feature that they are micron- to nano-sized particles that produce a US signal in response to an ultrasonic wave. Non-targeted UCA are currently used in echocardiography as injected microbubbles to opacify the blood pool and enhance endocardial border definition and facilitate more accurate assessment of left ventricular function in technically difficult echocardiograms [12]. The microbubbles are composed of perfluorocarbon or nitrogen gas encapsulated by shells of phospholipid, albumin, or biodegradable polymers, and remain within the intravascular compartment, thus requiring that the targeted molecules must be endoluminal. Because these gas bodies are much more compressible than water or tissue by several orders of magnitude, and are smaller than the wavelength of conventional diagnostic US, they undergo volumetric oscillation during imaging [4]. They have a proven clinical utility, particularly as a diagnostic tool in cardiology [38], radiology [7] and oncology [8]. Other UCA that are not gas-filled microspheres include liposomes [10] and liquid perfluorocarbon (PFC) emulsion nanoparticles that exit the microvasculture [27] and may allow acoustic targeting of extravascular markers because of their smaller size and longer circulating time [17]. Thus, ligand-coated (PFC) emulsion nanoparticles (250 nm in diameter) are used to identify the angioplasty-induced expression of tissue factor by smooth muscle cells within the tunica media [17]. Tissue factor-targeted nanoparticles bound to and increased the echogenicity of tissue factor expressing smooth muscle cells within the tunica media. The area of acoustic enhancement also appeared to coincide with the expression of induced tissue factor as revealed by immunohistochemistry.

The nanoparticles still have a small role in ultrasound imaging, for physical reasons. The frequencies necessary to detect particles smaller than 1 μm diameter, exceeding 30 MHz, do not penetrate more than a few millimetres into living tissues. The signal-to-noise ratio for ultrasound images is much lower with nanoparticles than those using microbubbles. Furthermore, contrast media are based on air–liquid interfaces of high curvature (bubbles), and creating bubbles of nanosize is a difficult undertaking [5].

Innovative methods to improve sensitivity and specificity are currently underway, such as delivering a higher payload of contrast agents to the target site and/or adding site-specific adhesion molecules to the shell, but little has been done towards optimization of the size of contrast agents. The targeting of UCA has been accomplished by the following strategies: 1) to select certain microbubble shell constituents that facilitate their attachment to cells in regions of disease [23, 24], 2) to attach disease-specific ligands such as monoclonal antibodies, peptides, and peptidomimetics to the microbubble shell surface [35].

As mentioned previously, the antigens that can be targeted by UCA are for the most part intravascular. A key concept for molecular imaging with contrast-enhanced ultrasound is that it requires attachment of a multivalent particle to target tissue in the face of vascular flow. Thus, the selection of a molecule to be targeted or the targeting ligand depends on many factors:

- bubble properties (ligand density; affinity and specificity; kinetics);
- hemodynamic properties (shear forces);
- target molecule properties (density and cellular distribution; specificity for disease; timing of expression; endogenous inhibitors).

The relative rate of unbound tracer clearance is also an important issue that determines temporal resolution. In this regard, with clearance time within minutes, microbubble tracers are ideal [3]. The density of adherent microbubbles retained in the vasculature at a target site has been shown to be extremely low with intra-vital microscopy, on the order of 10 microbubbles per mm^3 [32]. The targeting efficiency can be greatly increased with the application of ultrasound radiation force [31]. Imaging methods to sensitively detect and distinguish adherent targeted contrast agents from freely circulating agents and tissue are still unavailable. Thus, further investigations are needed. The potential sensitivity of ultrasonic molecular imaging could be weak due to the high background signal from free microbubbles and tissue obscured echoes from adherent microbubbles. Due to the lack of selectivity and sensitivity, current strategies for imaging adherent agents require waiting for clearance of freely flowing microbubbles and fragmentation of adherent agents [3]. A recent study has developed an innovative method based on a harmonic signal model of the returned echoes over a train of pulses, to emphasize the significance and potential of ultrasonic molecular imaging [42]. The detection techniques described by Zhao et al. are among the first examples of real-time methods to sensitively detect adherent targeted contrast agents based on their unique echo response, thus overcoming one of the main challenges with targeted imaging.

1.3 Proof of concept of molecular imaging with ultrasound

Molecular echography has shown promise for identifying angiogenesis. While the efficacy of therapeutic angiogenesis has been difficult to demonstrate in patients using traditional clinical metrics, such as exercise tolerance and nuclear single-photon emission CT (SPECT) imaging [34], ultrasound imaging of a molecular angiogenesis marker could provide a useful surrogate endpoint for a biological effect in patients undergoing therapeutic angiogenesis. Targeted UCA have been introduced to allow accurate visualization and quantification of molecular markers of angiogenesis in cardiology [22] and oncology [6]. Microbubbles targeted to alphavbeta3 ($\alpha v \beta 3$) integrins via peptides, [6, 20] or to vascular endothelial growth factor (VEGF) receptors via the non-heparin binding isoform of VEGF, VEGF121 [18], have been recently reported to identify angiogenic microcirculation in animal models of tumor-mediated or growth factor-mediated angiogenesis. A recent study has developed and validated a dual-targeted ultrasonographic imaging agent with microbubbles that attaches to both VEGFR2 and $\alpha v \beta 3$ integrins [41]. Such concepts can be extended to echographic imaging for myocardial and tumoral angiogenesis in response to therapeutic interventions, including genes or drugs.

In fact, the specificity of VEGFR2 and $\alpha v \beta 3$ integrin binding microbubbles and changes in marker expression during matrix metalloproteinase inhibitor treatment have been recently investigated. In tumors, accumulation of targeted microbubbles is significantly higher compared with non-specific ones and could be inhibited competitively by addition of the free ligand in excess [28]. Furtermore, the multimarker imaging approach could successfully be done during the same imaging session. Thus, targeted ultrasound appears feasible for the longitudinal molecular profiling of angiogenesis and for the sensitive assessment of therapy effects in vivo.

1.4 Targeted ultrasound contrast agents: new shuttles for drug delivery

The general goal of targeted drug delivery is to improve the efficacy of drug action in the region of the disease while reducing undesired side effects, such as toxicity, in healthy tissues. In order to improves the delivery of drugs and therapeutic genes, mechanical energy has been applied in the form of ultrasound irradiation. It is well known that ultrasound improves drug delivery into tissues and cells [14]. The growing interest concerning the engineering of targeted UCA has been also determined by the recent advances in gene therapy and molecular biology, and they represent an innovative approach in methods of non-invasive delivery of therapeutic agents. The well known application of gas-filled microbubbles, as contrast agent of diagnostic ultrasound [29, 33], has been combined as an effective technique for targeted delivery of drugs and genes [9].

The ultrasound-mediated microbubble destruction with co-administration of a drug or gene has been the first use of ultrasound-mediated drug delivery applications, and the effects remain spatially limited to the close proximity of the insonated microbubbles. Time limitations also exist: perturbations in a cell membrane caused by ultrasound-related effects are very short-lived. Therefore, the drug to be delivered should be located longer in the immediate vicinity of the insonated bubbles.

Consequently, the microbubbles themselves have been proposed as carrier vehicles for drugs and genes. The microbubbles can have drug molecules incorporated within the thick polymer shell or inside the gas core of thick-shelled bubbles. Drugs could be even attached to the external surface of the thin lipid monolayer of bubbles by covalent or non-covalent bonds, or incorporated into liposomes or nanoparticles that are then associated with the bubble surface. Thus, the effectiveness of microbubble-mediated drug delivery depends on following factors:

- shell composition;
- loading capacity;
- dose.

Microbubble shell composition can be easily adjusted by adding a variety of lipids to the emulsion mixture preparation or binding specific antibodies. The positively charged groups of some synthetic lipids make electrostatic interactions possible between the lipids and plasmid DNA, which possesses an overall negative charge. DNA coupled to the surface of the microbubbles remains intact even after insonation [2] and is also protected from enzymatic degradation [19]. However, in this configuration the loading capacity of the microbubbles is restricted to their surface area, and therefore, high amounts of microbubbles have to be injected to provide the desired effect. Unger et al. have demonstrated that it is possible to incorporate hydrophobic drugs into lipid microbubbles by simply adding them to the lipid phase prior to bubble preparation [36]. The use of monolayer-based lipid microbubbles is not the most beneficial for drug delivery of most hydrophobic pharmaceutical entities because the amount of drug carried in such a bubble preparation is quite limited. The thin shell even restricts the loading capacity, and it may not be able to prevent leakage of drugs from the bubble during their circulation through non-targeted areas of the body. One of the solving approaches is to prepare microbubbles with a thick lipid shell (i.e., polymeric microbubbles) containing triglyceride oil phase and a drug dissolved into it [37]; however, this approach is limited to hydrophobic fat-soluble drugs. Polymers of this kind are fully biocompatible, biodegradabile, and both hydrophobic and hydrophilic drugs can be incorporated.

The rate of drug release from the polymer microbubbles depends on the lipophilicity and water solubility of the drug. After destruction of the microbubble by US, the weak interactions between a hydrophilic drug and the polymer fragments will be broken easily, and the drug will be liberated rapidly. On the contrary, a hydrophobic drug will be released slowly, and gradually from the polymer fragments of destroyed microbubbles deposited in the tissue. The polymer shell can be filled with a perfluoropentane core, a liquid at room temperature that becomes gaseous at body temperature, to permit a faster release of the drug. It has been recently demonstrated that these particles upon insonation of tumors release the drug from the biodegradable polymer shell, which has resulted in tumor regression in an animal model [30].

If one wants to improve loading capacity of the bubbles, one could use multi-particle assemblies. Liposomes or nanoparticles that entrap the drug can be prepared separately and then coupled to the surface of the microbubbles [15]. When the microbubbles are destroyed by US in the target tissue, the energy released will cause rupture of the membrane of microbubble-associated liposomes and subsequently release the encapsulated drug [26]. However, these methods do not eliminate wash-out and systemic distribution of released drug by flowing blood.

More specific drug delivery may be achieved by attaching a ligand that is directed to a designated surface marker to the outside of the drug-laden microbubble. For example, endothelial surface markers are particularly attractive targets as certain markers are overexpressed in areas of angiogenesis, and targeted microbubbles have been shown to adhere to these markers [6]. Ultrasound can be applied locally to target bound microbubbles resulting in delivery of drug selectively in the areas where the surface marker is expressed.

The use of microbubbles as a tool for drug delivery enhancement has an enormous clinical potential, especially in oncology and cardiovascular applications. Whereas free drugs often possess harmful side effects, their encapsulation in microbubbles and subsequent local release, deposition, and potentiation in the target tissue by US triggering will help improve the therapeutic index, lower the incidence of adverse events, and achieve successful therapy. In fact, current tumor chemotherapy is associated with severe side effects caused by drug effects on healthy tissues. In addition, due to anomalous tumor vascularization and high interstitial pressure, spatial drug gradients are created in the tumor volume, resulting in survival of some cancer cells. These cells have a tendency to become drug resistant, thus dramatically decreasing the effect of subsequent treatment rounds. Drug encapsulation in targeted microbubbles has the potential to overcome both obstacles [13].

At least the ultrasound contrast microbubbles are relatively large (1–10 μm in diameter) compared with traditional pharmaceuticals. In some disease, such as tumors, the vessels are particularly permeable and often have large endothelial gaps [11]; anyway, contrast microbubbles are typically too large to exit the vasculature. This poses a particular problem when trying to target receptors that may be present in the tumor tissue rather than on the vascular endothelium. Thus, Wheatley et al. [40] have described a nanoparticle ultrasound contrast agent (450 nm diameter) with good acoustic properties.

Furthermore, the circulation time of ultrasound contrast agents is also a concern for ultrasound drug delivery, but it has increased in the last decade. The activation of the reticuloendothelial system may reduce the circulation time and not allow higher amounts of drug to be delivered to the target region. It is established that contrast agents are typically administered into a peripheral vein, so only a small amount of agent will pass through a tumor or injured tissue in a given circulatory cycle. Multiple circulations are necessary to allow destruction of enough agent to increase local concentration significantly. For example, polymer-shelled agents may provide a greatly increased circulation time [1].

Finally, the bioeffects associated with UCA are not fully understood and are strongly dependent on concentration and imaging strategies. These effects will need to be considered as clinical translation proceeds.

1.5 Conclusions

The future of molecular imaging by ultrasound contrast agents in tumor and injured tissues lies in multimodality and nanoparticle-based approaches, protein–protein interactions, and quantitative evaluation. Combinations of multiple modalities can yield complementary information and offer synergistic advantages over any modality alone.

Nanoparticles, possessing multifunctionality and enormous flexibility, can allow for the integration of therapeutic components, targeting ligands, and multimodality imaging labels into one entity, termed "nanomedicine," for which the ideal target is angiogenesis. Quantitative imaging of angiogenesis and protein–protein interactions that modulate angiogenesis will lead to more robust and effective monitoring of personalized molecular therapy.

UCA-mediated therapy as well as all other energy based treatment modalities will require imaging before treatment. Although this imaging can be performed in a variety of ways, the possibility for combining diagnostic and therapeutic ultrasound for imaging and treatment is especially attractive for reasons of simplicity and cost-effectiveness. Thus, some dual-modality imaging/therapy – focused ultrasound instruments should be considered.

Multidisciplinary approaches and cooperative efforts from many individuals, institutions, industries, and organizations are needed to quickly translate multimodality angiogenesis imaging into multiple facets of disease management, such as cancer. Not limited to cancer, these novel agents can also have broad applications for many other angiogenesis-related diseases.

Acknowledgements
This work was supported by the IST-FP6-EU Specific Targeted Research Project S.I.G.H.T. (systems for in-situ theranostics using micro-particles triggered by ultrasound) – contract number 033700.

References

1. Bloch SH, Wan M and Dayton PA et al. (2004) Optical observation of lipid- and polymer-shelled ultrasound microbubble contrast agents. Appl Phys Letts 84:631–3
2. Christiansen JP, French BA and Klibanov AL et al. (2003) Targeted tissue transfection with ultrasound destruction of plasmid-bearing cationic microbubbles. Ultrasound Med Biol 29:1759–1767
3. Christiansen JP and Lindner JR (2005) Molecular and Cellular Imaging with Targeted Contrast Ultrasound. Proceedings of the IEEE 93:809–818
4. Dayton PA, Chomas JE and Lum AF et al. (2001) Optical and acoustical dynamics of micro-bubble contrast agents inside neutrophils. J Biophys 80:1547–1556
5. Debbage P and Jaschke W (2008) Molecular imaging with nanoparticles: giant roles for dwarf actors. Histochem Cell Biol 130(5):845–875
6. Ellegala DB, Leong-Poi H and Carpenter JE et al. (2003) Imaging tumor angiogenesis with contrast ultrasound and microbubbles targeted to alpha(v)beta3. Circulation 108:336–341
7. Forsberg F, Liu JB and Merton DA et al. (1995) Parenchymal enhancement and tumor visualization using a new sonographic contrast agent. J Ultrasound Med 14:949–57
8. Goldberg BB, Raichlen JS and Forsberg F (2001) Ultrasound Contrast Agents. Martin Dunitz Ltd, London
9. Guo H, Leung JC and Chan LY et al. (2007) Ultrasound-contrast agent mediated naked gene delivery in the peritoneal cavità of adult rat. Gene Ther. 14(24):1712–1720
10. Hamilton AJ, Huang SL and Warnick D et al. (2004) Intravascular ultrasound molecular imaging of atheroma components in vivo. J Am Coll Cardiol 43:453–460
11. Hashizume H, Baluk P and Morikawa S et al. (2000) Openings between defective endothelial cells explain tumor vessel leakiness. Am J Pathol 156:1363–1380
12. Hundley WG, Kizilbash AM and Afridi I et al. (1998) Administration of an intravenous per-fluorocarbon contrast agent improves echocardiographic determination of left ventricular volumes and ejection fraction: Comparison with cine magnetic resonance imaging. J Am Coll Cardiol 32:1426–1432

13. Iyer A, Khaled G and Fang J et al. (2006) Exploiting the enhanced permeability and retention effect for tumor targeting. Drug Discov Today 11:812–818

14. Kassan DG, Lynch AM and Stiller MJ (1996) Physical enhancement of dermatologic drug delivery: iontophoresis and phonophoresis. J Am Acad Dermatol 34:657–666

15. Kheirolomoom A, Dayton PA and Lum AF et al. (2007) Acoustically-active microbubbles conjugated to liposomes: characterization of a proposed drug delivery vehicle. J Control Release 118:275–284

16. Klibanov AL (1999) Targeted delivery of gas-filled microspheres, contrast agents for ultrasound imaging. Adv Drug Deliv Rev 37:139–157

17. Lanza GM, Abendschein DR and Hall CS et al. (2000) Molecular imaging of stretch-induced tissue factor expression in carotid arteries with intravascular ultrasound. Invest Radiol. 35(4):227–234

18. Lee DJ, Lyshchik A and Huamani J et al. (2008) Relationship between retention of a vascular endothelial growth factor receptor 2 (VEGFR2)-targeted ultrasonographic contrast agent and the level of VEGFR2 expression in an in vivo breast cancer model. J Ultrasound Med 27(6): 855–866

19. Lentacker I, De Smedt SC and Demeester J et al. (2006) Microbubbles which bind and protect DNA against nucleases. J Control Release 116:e73–e75

20. Leong-Poi H, Christiansen J and Klibanov AL et al. (2003) Noninvasive assessment of angiogenesis by ultrasound and microbubbles targeted to alpha(v)-integrins. Circulation 107:455

21. Liang HD and Blomley MJK (2003) The role of ultrasound in molecular imaging. The British Journal of Radiology 76: S140–S150

22. Libby P (2002) Inflammation in atherosclerosis. Nature 420:868–874

23. Lindner JR, Coggins MP and Kaul S et al. (2000) Microbubble persistence in the microcirculation during ischemia/reperfusion and inflammation is caused by integrin- and complement-mediated adherence to activated leukocytes. Circulation 101(6):668–675

24. Lindner JR, Song J and Xu F et al. (2000) Noninvasive ultrasound imaging of inflammation using microbubbles targeted to activated leukocytes. Circulation 102:2745–2750

25. Lindner JR (2004) Molecular imaging with contrast ultrasound and targeted microbubbles. J Nuc Cardiol 11(2):215–221

26. Marmottant P and Hilgenfeldt S (2003) Controlled vesicle deformation and lysis by single oscillating bubbles. Nature 423:153–156

27. Marsh JN, Hall CS and Scott MJ et al. (2002) Improvements in the ultrasonic contrast of targeted perfluorocarbon nanoparticles using an acoustic transmission line model. IEEE Trans Ultrason Ferroelectr Freq Control 49:29–38

28. Palmowski M, Huppert J and Ladewig G et al. (2007) Multifunctional nanoparticles for combining ultrasonic tumor imaging and targeted chemotherapy. J Natl Cancer Inst 99:1095–1106

29. Price RJ, Skyba DM and Kaul S et al. (1998) Delivery of colloidal particles and red blood cells to tissue through microvessel ruptures created by targeted microbubble destruction with ultrasound. Circulation 98(13):1264–1267

30. Rapoport N, Gao Z and Kennedy A (2007) Multifunctional nanoparticles for combining ultrasonic tumor imaging and targeted chemotherapy. J Natl Cancer Inst 99(14):1095–1106

31. Rychak JJ, Klibanov AL and Hossack JA (2005) Acoustic radiation force enhances targeted delivery of ultrasound contrast microbubbles: in vitro verification. IEEE Trans Ultrason Ferroelectr Freq Control 52:421–433

32. Schumann PA, Christiansen JP and Quigley RM et al. (2002) Targeted-microbubble binding selectively to GPIIb IIIa receptors of platelet thrombi. Invest Radiol 37:587–593

33. Skyba DM, Price RJ and Linka AZ et al. (1998) Direct in vivo visualization of intravascular destruction of microbubbles by ultrasound and its local effects on tissue. Circulation 98(4):290–293

34. Simons M (2005) Angiogenesis: where do we stand now? Circulation 111:1556–1566

35. Takalkar AM, Klibanov AL and Rychak JJ et al. (2004)Binding and detachment of microbubbles targeted to P-selectin under controlled shear flow. J Control Release 96:473–482

36. Unger EC, McCreery TP and Sweitzer RH et al. (1998a) Acoustically active liposupheres containing paclitaxel: a new therapeutic ultrasound contrast agent. Invest Radiol 33:886–892

37. Unger EC, McCreery TP and Sweitzer RH (1998b) MRX-501: a novel ultrasound contrast agent with therapeutic properties. Acad Radiol 5 (1):S247–S249

38. Wei K and Kaul S (1997) Recent advances in myocardial contrast echocardiography. Curr Opin Cardiol 12:539

39. Wells PNT (2001) Physics and engineering: milestones in medicine. Med Eng Phys 23:147–153

40. Wheatley MA, Forsberg F and Oum K et al. (2006) Comparison of in vitro and in vivo acoustic response of a novel 50:50 PLGA contrast agent. Ultrasonics 44:360–367

41. Willmann JK, Lutz AM and Paulmurugan R et al. (2008) Dual-targeted contrast agent for US assessment of tumor angiogenesis in vivo. Radiology 248(3):936–944

42. Zhao S, Kruse DE and Ferrara KW et al. (2007) Selective imaging of adherent targeted ultra-sound contrast agents. Phys Med Biol 52(8):2055–2072

Chapter 2

Use of Microbubbles as Ultrasound Contrast Agents for Molecular Imaging

Mathieu Hauwel, Thierry Bettinger, and Eric Allémann

2.1 Introduction

The field of ultrasound contrast imaging has been literally bursting in the last decade. It becomes clear that ultrasonography which already plays a pivotal role in clinical diagnostics will become essential to health care management with the rise of ultrasound molecular imaging [6]. Technical breakthrough together with enhanced image processing greatly improved spatial resolution and dynamic range of ultrasound imaging. Ultrasound contrast agents (UCA) for medical imaging have been rapidly translated from exploratory research to clinical application [12]. The next generation of targeted contrast agents (TCA) is currently intensively investigated by numerous research groups. They will open a new era where diagnostics is not only based on pathophysiological signs but also on the molecular signature of a disease [2]. TCA are formulations generating signal enhancement, typically gas-filled microbubbles for ultrasound imaging, which are capable of specific binding to a site of interest. This is usually achieved by coupling a targeting ligand onto the surface of the microbubble. Three types of microbubbles are successively used for the different development stages of targeting projects: non-targeted naked microbubbles, streptavidin-bearing microbubbles and ligand-decorated microbubbles (Fig. 2.1).

Naked microbubbles could be used in vivo to set up imaging parameters, visualize tumor perfusion and perform dose ranging for optimal detection.

In a second phase, a targeted formulation is developed based on streptavidin-functionalized microbubbles, to which biotinylated ligands can be attached. This approach allows users to fine-tune their TCA. Indeed, by using this approach one can easily select the best targeting ligand among a large panel whilst optimizing its density at the bubble surface. These targeted microbubbles are also very convenient for preliminary in vivo experiments.

Finally, the definitive ligand is incorporated into the microbubble formulation by covalent coupling so that large batches of lyophilized TCA could be produced for pre-clinical evaluation.

Paradossi, G., Pellegretti, P., Trucco, A. (Eds.)
Ultrasound contrast agents. Targeting and processing methods for theranostics
© Springer-Verlag Italia, 2010

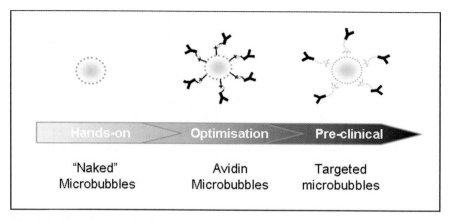

Fig. 2.1. Flowchart of the development of targeted microbubbles. Naked microbubbles are the ideal blank material to test both in vitro and in vivo experimental setups. Avidin-microbubbles are used for ligand density ranging in vitro and in vivo. Finally pre-clinical grade targeted microbubbles produced in large batches by covalent coupling are meant for large-scale in vivo studies in several animal models

2.2 Molecular design

2.2.1 Contrast media

Two classes of ultrasound contrast media are prepared at Bracco Research: blood pool agents and targeted agents. The former circulate in the blood stream for a given period of time revealing organ perfusion whereas the latter accumulate specifically in the vasculature of a target organ. Both types could be composed of either soft-shell or hard-shell microparticles.

Soft-shell microparticles are best illustrated by the commercial contrast agent SonoVue®. It is composed of insoluble gas (SF_6) microbubbles, dispersed in an aqueous medium and stabilized by a phospholipid monolayer [11] (Fig. 2.2). Sulphur hexafluoride (SF_6) and perfluorocarbon gases, such as perfluorobutane (C_4F_{10}) and perfluoropropane (C_3F_8), are appropriate candidates because they are inert and non-toxic with a very low solubility in aqueous media generating stable microbubbles. Sheathing these microbubbles with biocompatible phospholipids not only increases their plasma half-life even further, but also narrows their size distribution and preserves their physical integrity. The soft-shell agents can be lyophilized for long-term storage while retaining the same acoustical and biochemical characteristics upon reconstitution. The contrast agents injected intravenously persist long enough in the blood stream to perform ultrasonographic scans. After a few minutes, the fluorinated gas is exhaled and the components of the soft-shell are metabolized.

Hard-shell agents are gas-filled microcapsules made of synthetic polymers or lipids. Because of the stiffness of their shell they are much more resistant than bubbles to mechanical stress but also less echogenic. In many cases,

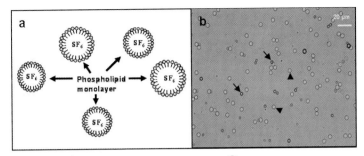

Fig. 2.2. SonoVue® microbubbles. (**a**) SonoVue® is composed of phospholipidic shells entrapping sulphur hexafluoride microbubbles. (**b**) SonoVue® and targeted microbubbles have an average diameter of 1.5 µm and a narrow size range. All bubbles (arrows) are smaller than red blood cells i.e. 8 µm (arrow heads)

detection can only be achieved by using ultrasound pulses that burst microcapsules. This atypical behavior however can be harnessed for different applications such as gene and drug delivery by sonoporation [5, 8] and to improve thrombolytic procedures [10].

Targeted contrast agents developed for molecular imaging are generally based on soft-shell technology as they are used in a non-destructive real-time contrast imaging mode.

2.2.2 Targeting ligands

Targeting moieties such as antibodies, antibody fragments, peptides or small chemical compounds, can be coupled to the surface of microbubbles. There is a broad variety of approaches described in the literature for coupling or attaching ligands to structures containing phospholipids [1, 4]. For the TAMIRUT project, antibodies directed against tumor markers were chosen as targeting ligands and attached to microbubbles by two different methods: in a first approach, the strong binding affinity of streptavidin for biotin was exploited for the preparation of targeted microbubbles. Then, microbubbles with a covalent chemical coupling of ligands were formulated. Targeted contrast agents were prepared first by incorporating streptavidin molecules in the composition of microbubbles by thiol-maleimide coupling chemistry. On the other hand, target specific antibodies can be biotinylated using standard protocols when not available commercially (Fig. 2.3a). Following resuspension of the lyophilized cake with physiological saline solution, biotinylated antibodies at any desired concentration are added. After a short incubation at room temperature, all the antibodies are captured at the microbubble surface (Fig. 2.3b). The contrast agent is then ready for injection. Using this approach, antibody density at the bubble surface can be easily modulated. Alternatively, antibodies can be directly attached to the microbubble surface by covalent coupling using heterobifunctional linkers (Fig. 2.4a) prior to lyophilization, in which

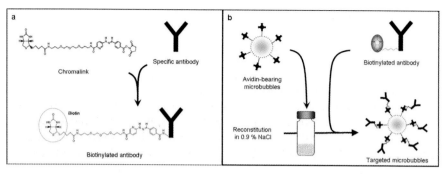

Fig. 2.3. Steptavidin-microbubbles. (**a**) Targeting antibodies are readily biotiny-lated on amine residues using NHS-activated biotin reagents such as Chromalink Biotin 354S (Solulink). After a single purification step, the number of biotin per antibody molecule is quantified by spectrophotometry. (**b**) Streptavidin, a tetrava-lent biotin-binding protein is added to the surface of microbubbles by chemical cou-pling to phospholipids via a polyethylene glycol spacer. Biotinylated antibodies and streptavidin-microbubble suspension are mixed together and incubated for 10 min at room temperature under agitation. Antibody density, microbubble concentration and size distribution are measured before use

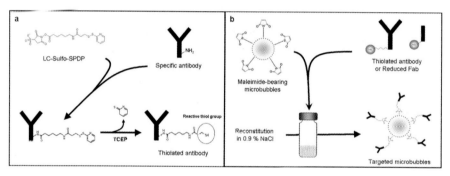

Fig. 2.4. Targeted microbubbles. (**a**) Targeting antibodies are functionalized with heterobifunctional linkers, such as LC-sulfo-SPDP (Pierce), introducing a protected thiol group. The reactive thiol group is unprotected by reduction with TCEP imme-diately before coupling. The number of thiol groups per antibody molecule is quanti-fied by titration of pyridine-2-thione release. (**b**) Whole antibodies or Fab with reac-tive thiol groups and microbubbles bearing maleimide functions are mixed together then lyophilized. Ligand density, microbubble concentration and size distribution are analyzed on resuspended bubble solution

case the contrast agent can be used immediately upon reconstitution of the vial (Fig. 2.4b). Covalent coupling is particularly attractive because a vari-ety of chemical groups present on the targeting ligand can be exploited for coupling chemistry. Besides, this approach is well suited for large scale pro-duction. In addition, since antibody density is set during the manufacturing process, all vials of a batch have the same targeting properties.

2.3 In vitro screening

2.3.1 Dynamic binding assay

In order to select the best targeting ligand, binding abilities of TCA have to be evaluated in conditions mimicking as closely as possible those found in the blood stream. For instance, shear stress corresponding to heamodynamic conditions can be simulated in a laminar flow chamber placed under a microscope (Fig. 2.5a). For this, the molecular target to be tested is first coated on the top of the flow chamber (Fig. 2.5b). Then, targeted microbubbles are infused through the chamber at a controlled flow rate using a precision infusion pump. Composition and viscosity of the suspension is adjusted using a mixture of buffer and human plasma and flow parameters are set to simulate venous circulation (i.e. wall shear stress $= 2$ dynes/cm^2 and wall shear rate $= 114\,s^{-1}$ [3]) where most microbubbles accumulate in vivo. Bright field images are acquired every 2 min for 10 min. Then, microbubbles are counted and accumulation rate in the field of view (185×138 µm) is expressed in number of bound bubbles per minute.

2.3.2 Influence of ligand density on binding efficiency

Microbubbles targeting the Vascular Endothelium Growth Factor-Receptor 2 (VEGFR2), a tumor angiogenesis marker, were synthesized with four differ-

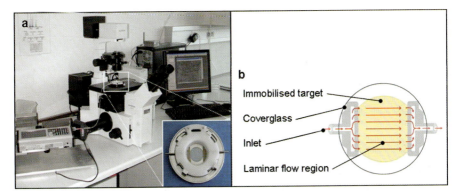

Fig. 2.5. Parallel flow chamber set up. (**a**) The parallel flow chamber is placed upside down under an inverted microscope so that the coated surface is on the top. This configuration favors the binding of buoyant particles such as microbubbles. The bubble suspension is loaded into a 50 ml syringe and perfused through the flow chamber using a precision pump. Images are acquired with a CCD camera connected to a computer. (**b**) The molecular target is immobilized on a glass coverslip (yellow area) placed over a micro-aqueduct glass slide having two "T" shaped grooves on each side (grey shape). The laminar fluid pathway (red arrows) is created in the space between the two glass slides. Both grooves are connected to perfusion holes for microbubble suspension inlet and waste collection

ent ligand densities and tested in dynamic flow conditions on recombinant human VEGFR2 coating. VEGFR2 density on the glass slide was set to 60 molecules/μm^2, which is two to four times lower than the estimated physiological density [7], in order to have stringent binding conditions. Microbubbles with as little as 40 antibodies per square micrometer (i.e. 363 antibodies per bubble) were able to bind to the glass slide demonstrating a very good sensitivity (Fig. 2.6a). Microbubble accumulation rate then rapidly increases with ligand density reaching a plateau near 2'300 antibody/μm^2 (Fig. 2.6b). It must be stressed that this maximal accumulation rate did not result from the saturation of all binding sites as only a few percent of the surface is covered by microbubbles. This plateau rather reflects the binding potential of a given ligand in flow condition, i.e. its statistical chance to bind to the target. Although this dynamic binding assay is a faithful model of the environment encountered by microbubbles in the blood flow, it cannot be used to establish an optimal ligand density. The microbubble formulations need to be evaluated when faced with the complexity of a relevant animal model.

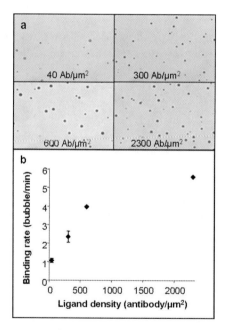

Fig. 2.6. In vitro screening of VEGF Receptor 2-specific microbubbles. (**a**) Targeted bubbles with as low as 40 antibodies per square micrometer were able to bind to a recombinant VEGFR2-coated coverglass. Bubble accumulation was directly dependant on ligand density on the surface. (**b**) Quantitative analysis of microbubble accumulation showed that the binding rate increases with ligand density before reaching a plateau

2.4 In vivo proof of concept

2.4.1 Methodology

Most clinical ultrasound systems feature contrast specific modes taking advantage of the non-linear acoustic response of microbubbles. The high frequency ultrasound scanner Vevo770 from Visualsonics, which is dedicated for small animal echography imaging, has no particular detection capabilities for contrast agents but offers a software alternative that allows microbubble signal extraction from fundamental B-mode cineloops. First, a tissue signal dataset is created by acquiring images before injection of the contrast agent. Then, these references are subtracted from contrast-enhanced frames to extract raw microbubble signals. Finally, microbbubble signals and tissue signals are merged together resulting in a colorized contrast overlay image (Fig. 2.7). Even after contrast agent signal extraction, images usually contain a great deal of artefact, some of which is due to circulating bubbles and speckles. A physical subtraction method, in this case, is used to get rid of remaining background noise before quantitative analysis. To this aim, all bubbles in the field are destroyed using a short high power pulse. Signal recorded shortly after bubble destruction is only background noise, consequently the signal difference before and after bubble destruction can be regarded as the specific contribution of targeted microbubbles (Fig. 2.8). Non-destructive imaging is performed at low acoustic power. Raising the acoustical power results in microbubble destruction and clearance of all bubbles in the field. Accordingly, signal intensity is high before destruction due to microbubble echoes and is reduced to background noise after destruction. Further destruction pulses have no effects on background noise.

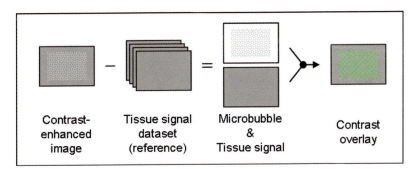

Fig. 2.7. Microbubble signal extraction from fundamental B-mode. Microbubble signals are separated from anatomical information by subtraction of a tissue signal reference acquired beforehand to the contrast-enhanced image. Keeping the ultrasound probe still with a probe holder is an absolute requirement to generate an accurate tissue signal dataset. Microbubble signal is colorized and merged with tissue signal to produced a so-called contrast overlay image

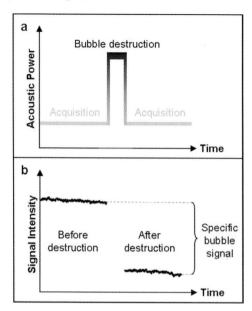

Fig. 2.8. Background subtraction by bubble destruction. (**a**) Microbubbles are destabilized by high acoustic power ultrasound waves and eventually collapse. For this reason, acquisition is performed at non-destructive power. Raising the acoustical power results in microbubble destruction and clearance of all bubbles in the field. (**b**) Accordingly, signal intensity is high before destruction due to microbubble echoes and is reduced to background noise after destruction, thus the difference between pre- and post-destruction signals is regarded as the specific bubble contribution. Further destruction pulses have no effects on background noise.

2.4.2 Tumor detection

Microbubbles targeted against VEGFR2 or control microbubbles (functionalized with an isotype-control antibody) were injected intravenously via the tail vein of Balb/c mice bearing subcutaneous DA3 mammary tumors. Control and targeted microbubbles were injected sequentially, in randomized order, to each mouse within the experimental group, ensuring for obvious reasons that the entire signal from previous injections had completely disappeared. The contrast agent "wash-in" phase was followed for 45 seconds post injection. Microbubble accumulation was imaged after 10 min using the acquisition-destruction-acquisition sequence described above and specific microbubble signals were quantified in a region of interest drawn around the tumor. As expected, signal intensity was much higher in the tumor when mice received targeted bubbles (Fig. 2.9) compared to control bubbles (Fig. 2.10). After destruction, signal intensities in both cases were comparable, which shows that background noise levels are not contingent on the nature of the bubbles injected. This resulted in a five fold higher signal difference with VEGFR2 tar-

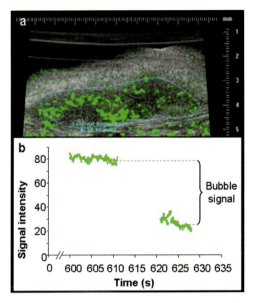

Fig. 2.9. Tumor detection with microbubbles specific for VEGFR2. (**a**) Signal was clearly visible in the tumor when mice received VEGFR2-targeted microbubbles although the spots were heterogeneously distributed. (**b**) Absolute signal intensity dropped from 80 to 25 following destruction (i.e. a net difference of 55)

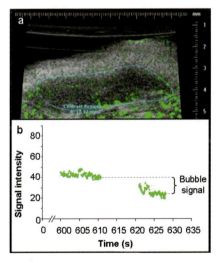

Fig. 2.10. Tumor detection with control microbubbles. (**a**) Very few non-targeted microbubbles accumulated in the tumor (delineated by a blue line) 10 min after injection. (**b**) A 10 seconds cineloop was recorded 10 min post-injection of microbubbles. Then, a 10 seconds destructive pulse was applied immediately followed by another 8 seconds cineloop. Absolute signal intensity was only 40 before destruction and 25 after (i.e. a net difference of 15)

geted microbubbles than with control microbubbles. A 10 seconds cineloop was recorded 10 min post-injection of microbubbles. Then, a 10 seconds destructive pulse was applied immediately followed by another 8 seconds cineloop. (a) When mice received control (non-targeted) microbubbles, very few accumulated in the tumor after 10 min. Absolute signal intensity was only 40 before destruction and 25 after (i.e. a net difference of 15). (b) Signal was clearly visible in the tumor when mice received VEGFR2-targeted microbubbles although the spots were heterogeneously distributed. Absolute signal intensity dropped from 80 to 25 following destruction (i.e. a net difference of 55).

2.5 Conclusions and prospects

A full Targeted contrast agent (TCA) development platform has been set up to perform molecular imaging of vascular targets. Targeting ligands directed towards cancer markers were coupled to microbubbles by two different approaches allowing to fine tune their binding performance. Refined TCA, once validated in vitro, can be used in vivo in animal models. Thus, TCA specific for VEGFR2 for instance, were successfully used to image subcutaneous mammary tumors in mice. Because they are able to bind tightly to their target, targeted microbubbles generate a typical signal enhancement of the tumor compared to non-targeted microbubbles. As shown in vitro [9], the number of bound bubbles can be correlated to the target density. Quantitation of tumor marker density in vivo using ultrasound biosensors would be a tremendous breakthrough for non-invasive cancer grading and staging. There are however several technical hurdles that remain to be overcome such as the variability of signal attenuation among patients, the unknown size distribution of bound bubbles and the binding tropism of TCA within the different vascular compartments. These issues were tackled by other members of the TAMIRUT consortium including groups at the University of Twente, University of Genoa and Esaote whose results are presented in other chapters of the present book.

Acknowledgements
The authors would like to acknowledge Dr T. Messager, P. Bussat, A. Helbert and S. Pagnod-Rossiaux for their contribution to the work presented herein.

References

1. Hermanson GT (1996) Bioconjugate techniques, ed. A. Press, Elsevier
2. Kaufmann BA and Lindner JR (2007) Molecular imaging with targeted contrast ultrasound. Curr Opin Biotechnol 18(1):11–16
3. Kim MB and Sarelius IH (2003) Distributions of wall shear stress in venular convergences of mouse cremaster muscle. Microcirculation 10(2):167–78

4. Klibanov AL (2005) Ligand-carrying gas-filled microbubbles: ultrasound contrast agents for targeted molecular imaging. Bioconjug Chem 16(1):9–17
5. Klibanov AL (2006) Microbubble contrast agents: targeted ultrasound imaging and ultrasound-assisted drug-delivery applications. Invest Radiol 41(3):354–362
6. Lindner JR (2002) Evolving applications for contrast ultrasound. Am J Cardiol 90(10A):72J–80J
7. Mac Gabhann F and Popel AS (2004) Model of competitive binding of vascular endothelial growth factor and placental growth factor to VEGF receptors on endothelial cells. Am J Physiol Heart Circ Physiol 286(1):H153–164
8. Newman CM and Bettinger T (2007) Gene therapy progress and prospects: ultrasound for gene transfer. Gene Ther 14(6):465–475
9. Ottoboni S et al. (2006) Characterization of the in vitro adherence behavior of ultrasound responsive double-shelled microspheres targeted to cellular adhesion molecules. Contrast Media Mol Imaging 1(6):279–290
10. Porter TR and Xie F (2001) Ultrasound, microbubbles, and thrombolysis. Prog Cardiovasc Dis 44(2):101–110
11. Schneider M et al. (1995) BR1: a new ultrasonographic contrast agent based on sulfur hexafluoride-filled microbubbles. Invest Radiol 30(8):451–457
12. Wilson SR and Burns PN (2006) Microbubble contrast for radiological imaging: 2. Applications. Ultrasound Q 22(1):15–18

Chapter 3

Design of Novel Polymer Shelled Ultrasound Contrast Agents: Towards an Ultrasound Triggered Drug Delivery

Mariarosaria Tortora, Letizia Oddo, Silvia Margheritelli, and Gaio Paradossi

3.1 Introduction

Ultrasound contrast agents (UCA) have been used for years in the clinical field, for applications such as blood pool enhancement, characterization of liver lesions or perfusion imaging [12–15, 20]. The contrast agents are generally in the form of spherical voids or cavities filled by a gas, called microbubbles (MB). MBs are stabilized by a coating material such as phospholipids, surfactants, denatured human serum albumin or synthetic polymers. As gas is less dense than liquids or solids, sound travels more slowly in gas than it does in liquid. The difference in the sound speed in the microbubbles creates an acoustic mismatch between tissue and blood surrounding a microbubble, making it an efficient reflector of ultrasound energy. The ability to produce strong signals from microbubbles depends on the stability of the gas particle. Stability has been enhanced by varying the gas composition (air, nitrogen, sulphur hexafluoride, perfluorocarbons) and/or by chemical modification of the microbubble shell [20]. As microbubbles stay in the vascular space and have a behavior similar to red blood cells in the microcirculation, they can be used as intravascular tracers. The possibility to introduce a targeting ligand for a specific interaction with its receptor onto the microbubble shell can be a way to selectively accumulate the contrast agent in the deseased region. Moreover, targeted MBs can be used to carry different drugs to the desired sites and release them after their destruction/cavitation with ultrasound radiation or by chemical/enzymatic cleavage.

In Table 3.1, the main properties of an ideal microbubble for diagnostic and therapeutic purposes are reported. In order to be injectable, an average external diameter lower than 5 μm and a narrow size distribution are mandatory. The ultrasound scattering efficiency is related to the difference in density and compressibility between medium and microbubbles. But among all features, the presence of suitable reactive chemical functionalities at the surface, for targeting and loading purposes, are really important for an active contrast agent. Finally, the circulation lifetime is an important aspect, because

Paradossi, G., Pellegretti, P., Trucco, A. (Eds.)
Ultrasound contrast agents. Targeting and processing methods for theranostics
© Springer-Verlag Italia, 2010

Table 3.1. Features of an "ideal" ultrasound contrast agent

Functional property	*Structural property*
Injectability	Average external diameter $< 5\mu m$ Narrow size distribution
Ultrasound scattering efficiency	Highest density and compressibility difference between medium and microbubbles
Biocompatibility	Suitable surface chemical moieties
Drug payload	Suitable reactive chemical functionalities at the surface
Drug and/or gas delivery	collapsing within a mechanical index (MI) < 0.9 Narrow thickness distribution

the particles must have enough time to release the drug in the appropriate regions.

Previous studies explored the use of the conventional lipid shelled microbubbles as targeted microbubbles for ultrasound imaging. These devices have several drawbacks to general use, including high dimensional variability and short circulation time and stability in the bloodstream. Additionally, the shell of a lipid coated microbubble is not an ideal drug reservoir because its molecular thickness and surface area allows only a limited loading capacity. Finally, the field of drugs that can be used is restricted to amphiphilic or charged molecules. The addition of drug payload and drug delivery to the properties of microbubbles can change the nature and function of the ultrasound contrast agents.

Our research is focussed on developing a new concept of ultrasound contrast agent (UCA), consisting of a multifunctional device able to behave as a diagnostic tool, providing high quality images and as a therapeutic agent, for in situ delivery. The concept of next-generation UCA incorporates in its operating mode the main features of theranostics. In the context, we will describe the fabrication of this new class of microbubbles using poly(vinyl alcohol), PVA, as a starting material, a synthetic polymer with good biocompatibility properties. Hereafter we will call these air filled, polymer shelled microdevices "microballoons", following the common habit of the literature.

The process for obtaining these particles originates from foaming a modified PVA solution where a crosslinking reaction takes place at the water/air interface. Chemical modifications of the outside surface of the MB polymeric shell open a wide range of possibilities for drug loading and specific targeting. A critical discussion of the synthetic strategies used for the surface modification of MBs is reported in this chapter. In vitro biocompatibility assessment of some of the functionalized MBs have been carried out and reported in the chapter "Polymer based biointerfaces: a case study on devices for theranostics and tissue engineering."

3.2 Experimental section

3.2.1 Materials

Poly (vinyl alcohol) (PVA) with a number average molecular weight (Mn) of 35000, 4-Nitrophenyl chloroformate (NPC), 4-dimethylamino pyridine (DMAP), dimethylsulfoxide (DMSO) anhydrous, 1-Methyl-2-pyrrolidinone (NMP) anhydrous and pyridine anhydrous were purchased from Sigma.

Arginine-Glycine-Aspartic acid, Arg-Gly-Asp (RGD, MW 346.3) was obtained from Espikem Srl, Florence, Italy.

O,O'-Bis(2-aminoethyl)polyethylene glycol (NH_2-PEG-NH_2) with a molecular weight (Mw) of 2000, L-cysteine hydrochloride monohydrate, and γ-cyclodextrin (γCD) were Fluka products. The Micro BCA^{TM} (bicinchonic acid) Protein Assay Kit was purchased from Pierce Chemical Co. All inorganic reagents were Carlo Erba products. Reagents were used as received.

Water was Milli-Q purity grade with resistivity of 18.2 M$\Omega \cdot$ cm with a deionization apparatus (Pure Lab) from USF.

3.2.2 Methods

Synthesis of PVA microballoons at pH5 and room temperature (MB5R)

2 g of PVA and 100 ml of Milli-Q water were added in a 250 ml beaker. The suspension was stirred at 80°C until complete PVA dissolution. 0.19 g of sodium metaperiodate was added to the solution, and the reaction was carried out at 80°C for 1 h with constant stirring. After cooling the solution to room temperature, the cross-linking reaction proceeded with vigorous stirring with an Ultra-Turrax T-25 at 8000 rpm equipped with a Teflon-coated tip at room temperature for 2 hours. Floating microballoons were separated from solid debris and extensively washed in separatory funnels and stored in Milli-Q water for further use.

Synthesis of PVA microballoons at pH 5 and 5°C (MB5C)

MB5C microballoons were obtained by following the same procedure outlined above but keeping the beaker containing the telechelic PVA solution in an ice-water bath during the cross-linking reaction.

Synthesis of PVA microballoons at pH 2 and 5°C (MB2C)

MB2C microballoons were obtained by following the same procedure used for the preparation of MB5C, but adding 1 ml of H_2SO_4 1M into the reaction medium before the cross-linking reaction. After 2 hours, the medium was neutralized with NaOH 0.1M.

Functionalization of MBs with NH$_2$-PEG-NH$_2$

12 mg of NH$_2$-PEG-NH$_2$ was dissolved in 2 ml of water and added to 20 ml of MB suspension (1 mg/ml). Then 2 mg of NaBH$_3$CN was added to the suspension and the pH was adjusted to 5. After 3 days at room temperature, microballoons were extensively dialyzed to remove unreacted NH$_2$-PEG-NH$_2$.

Functionalization of MBs with RGD

2 ml of a 0.01M RGD solution in water was added to a 20 ml MB suspension (1 mg/ml). Then 2 mg of NaBH$_3$CN was added to the suspension and the pH was adjusted to 5. After 3 days at room temperature, microballoons were extensively dialyzed to remove unreacted RGD. The amount of RGD present on MBs was determined by BCA colorimetric assay [3].

Functionalization of MBs with L-cysteine(Cys-MB)

1 g of L-cysteine hydrochloride monohydrate was added to 20 ml of MB suspension (2 mg/ml). Then 72 mg of NaBH$_3$CN was added to the suspension and the pH was adjusted to 5. After 3 days at room temperature, microballoons were extensively dialyzed to remove unreacted cysteine.

Thiol content introduced to the microbubbles by functionalization with L-cysteine was determined by measuring the absorbance at 412 nm on centrifuged solutions, after reaction of the MBs with Ellman's reagent for 15 min at room temperature.

Synthesis of γ- cyclodextrin (γCD) – functionalized MBs

For γCD modification, MBs prepared at pH 5 and room temperature were used. γCD was bound to the MBs via acetalization reaction, taking advantage of the large number of hydroxylic groups present on the MB surface. A 0.03 M γCD solution was prepared and an amount of sodium metaperiodate sufficient to oxidize one sugar unit per ring was added. Oxidation proceeded overnight at room temperature. An equal volume of MB suspension (1 mg/ml) was then added to the solution and the pH adjusted to 3 with HCl. After 24 h, the γCD-MB suspension was extensively washed. To quantify the bound cyclodextrin, γCD-MBs were then titrated with phenolphthalein, measuring the solution absorbance at 550 nm.

Loading of NO on Cys-MB (NO-Cys-MB)

10 ml of a 1 mg/ml suspension of Cys-MB was centrifuged. The recovered Cys-MBs were suspended in 5 ml of HCl 0.1M, then 54 μl of a 0.1M sodium nitrite solution was added. The reaction was conducted at 37°C for 30 minutes. The functionalized MBs were washed three times with HCl 0.1M, twice with water and then suspended in PBS solution. The presence of NO on MB

aqueous dispersions was monitored by electron paramagnetic resonance (EPR) spectroscopy an EMX Bruker spectrometer (Bruker Germany) operating at T = 100 K at 0.5 mT field modulation.

Synthesis of NPC functionalized PVA (NPC-PVA)

1 g of PVA and 45 ml of NMP were poured into a two necked flask. The suspension was stirred at room temperature until complete PVA dissolution. 8 ml of pyridine and 0.4 g of DMAP were added. Then 4 g of NPC was introduced; during the NPC addition, the flask was kept in an ice-water bath. After 24 hours, the solution was transferred into a beaker and 150 ml of ethanol was added, allowing the polymer to precipitate. The white precipitate was collected and washed with an excess of ethanol and finally dried overnight.

Synthesis of RGD functionalized PVA (RGD-PVA)

100 mg of NPC-PVA was introduced into a 50 ml flask and dissolved in 8 ml of DMSO at 60°C. After cooling the solution to room temperature, 0.8 ml of pyridine and 15 mg of DMAP were added, then 0.65 ml of a 0.2 M solution of RGD in water was dropped into the flask. After 3 days at room temperature, the reaction mixture was dialyzed against a solution of NaOH at pH 11 and then with Milli-Q water. A white precipitate was obtained after freeze drying. The amount of RGD grafted to the PVA chains was estimated by H1-NMR spectroscopy.

Synthesis of telechelic PVA

1 g of PVA and 8.3 ml of Milli-Q water were added to a 20 ml beaker. The suspension was stirred at 80°C until complete PVA dissolution. Approximately 0.095 g of sodium metaperiodate was added to the solution and the reaction was carried out at 80°C for 1 h with constant stirring [8].

Synthesis of telechelic PVA hydrogels

To 8.3 ml of the telechelic PVA solution, 1.5 ml of HCl 3M was added and the cross-linking reaction proceeded at room temperature for 24 hours. Hydrogels were washed with Milli-Q water to pH 5.

Synthesis of RGD-PVA hydrogels

54 mg of RGD-PVA powder was suspended in 3 ml of water and the solution was transferred into a 10 ml beaker. The temperature was increased to 80°C, then 5 mg of $NaIO_4$ was added and the reaction was performed for 1 h with constant stirring. After cooling the solution to room temperature, 8.3 ml of telechelic PVA solution and 0.5 ml of HCl 12 M were added. The cross-linking reaction proceeded for 24 hours, then the hydrogels were extensively washed with Milli-Q water to remove the excess of HCl.

3.3 Results and discussion

3.3.1 Synthesis of PVA microballoons and surface characterization

The fabrication of PVA based MBs has been already reported [6] and it consists of a two steps procedure. The first step of the reaction is based on a selective metaperiodate oxidation of the head-to-head sequences containing the hydroxylic groups present in the backbone of PVA. The product of this splitting reaction is a telechelic PVA with an aldehydic group at each chain end. These PVA macromers are then cross-linked by an acetalization reaction at the water/air interface under high shear stirring. The two reaction steps are reported in Scheme 1.

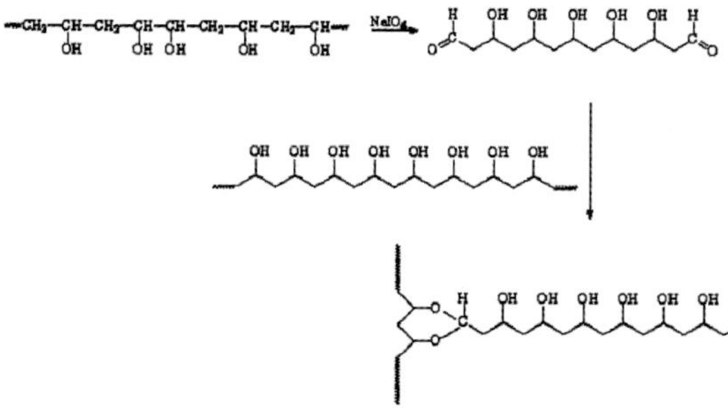

Scheme 1. Periodate splitting of PVA vicinal diols and acetalization for crosslinking reaction

The yield of microballoons depends on several experimental parameters and the choice of proper operating conditions is key to a successful microballoon design. In particular, temperature and pH are very important features in determining their diameter, size and shell thickness. Three kinds of MBs were prepared: MB5R, obtained at pH 5 and room temperature, MB5C, obtained at pH 5 and 5°C and MB2C, obtained at pH 2 and 5°C. The size and the shell thickness were determined by confocal laser scanning microscopy (CLSM) on rhodamine-labeled microballoons; a picture of MB labelled with rhodamine is reported in Figure 3.1. In the insert the intensity profile used to evaluate the shell thickness is shown. The MB features of the three types of MBs are summarized in Table 3.2. CLSM is a handy and valuable tool for the structural investigation of mesoscopic particles. However, the resolution of CLSM (about 0.5 μm) allows only a rough estimation of the bubble shell thickness. In the chapter "Novel Characterization Techniques of Microballoons" of this book, more accurate approaches for a structural investigation of polymeric microbubbles are described.

Freeze fracture electron microscopy analysis offer a valuable investigation tool as it allows a transverse view of the microballoons. The electron microscopy revealed fibrillar morphology with an outer region characterized

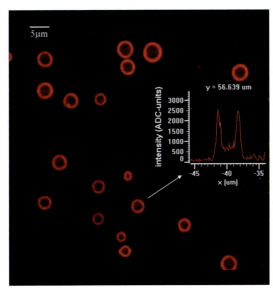

Fig. 3.1. Laser scanning confocal micrograph of PVA shelled microballoons labelled with rhodamine. Insert: Fluorescence intensity profile used to evaluate the shell thickness

Table 3.2. Microballoon structural features in different synthesis conditions

MB type	External diameter (μm)	Shell thickness (μm)
MBpH5R	4.0 ± 0.8	0.6 ± 0.1
MBpH5C	2.0 ± 0.3	0.4 ± 0.1
MBpH2C	2.4 ± 0.4	0.6 ± 0.1

by loosely arranged PVA fibrils and an inner region where the polymeric material is more compact. In the outer domains of the shell, PVA fibrils protrude radially toward the outside. By collecting the information, the hypothetic MB structure can be described as a cavity filled by air and a shell formed by the cross-linked PVA chains extending into the solution and forming a "hairy" surface (see Fig. 3.2).

3.3.2 Functionalization of MBs

Derivatization and drug loading of PVA microballoons depend on the reactivity of different functional groups on their surface. Moreover, aqueous medium is required for the coupling of molecules to the MB surface. Taking into account these restrictions, acetalization and reductive amination are the two main synthetic routes used to conjugate functional molecules to the MB surface. Molecules carrying aldehyde can be conjugated by acetalization of the

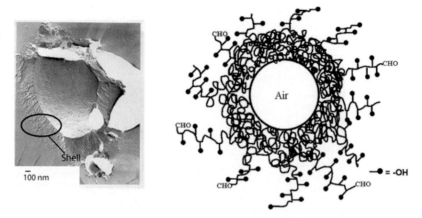

Fig. 3.2. Freeze fracture electron micrograph of a PVA shelled microballoon (left) and a hypothetical description of the MB PVA shell (right)

hydroxyl groups of the MB surface. On the other hand, during MB shell formation, unreacted aldehyde end groups forming intramolecular hemiacetals are still present on the MB surface in the form of hemiacetals. These groups can be used to functionalize the MB surface by reductive amination of molecules carrying amino groups or by acetalization reaction with molecules carrying hydroxyl groups.

Parenteral administration of particles involves contact with biological fluids. After injection, these microdevices could become coated with serum proteins, with potential biological effects. In fact, the close spatial packing of the adsorbed proteins can lead to exposure of novel epitopes, perturbed function and/or avidity effects. Moreover, phagocytosis of microparticles is vastly affected by the size and shape of the particles as well as by the surface features of the microparticles [19]. PEG stealth properties are known [9]. It is a polymer used as surface coating to prevent protein adsorption onto the particles, their recognition and degradation.

According to the syntehtic strategies outlined above, pegylated MBs were obtained by coupling the unreacted hemiacetals present on the MB surface with amine end groups of diammino poly(ethylene glycol) chains, NH_2-PEG-NH_2. The aldehyde groups masked as hemiacetals on MBs were activated at low pH. NH_2-PEG-NH_2 was added to the suspension to perform Schiff base coupling followed by reductive amination with sodium cyanoborohydride. The aqueous suspension of MBs was exhaustively dialyzed against water. The MB suspension was freeze dried and the dried powder was dispersed in D_2O for ^{1}H-NMR analysis. As shown in Figure 3.3, the presence of PEG chains grafted onto the MB surface was evidenced by the resonance centered at 3.7 ppm attributed to the protons of methylene groups in the PEG chains. The peaks at 4.1 ppm and 1.5 ppm were attributed to the PVA backbone methine and methylene groups [2], respectively.

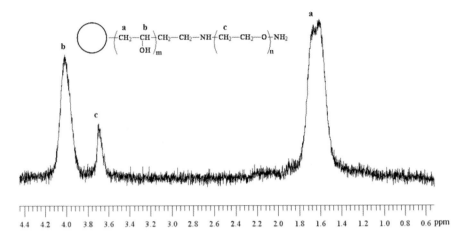

Fig. 3.3. H1-NMR spectrum of pegylated PVA shelled MBs in D_2O

In order to introduce a biospecific activity to the MBs and in particular to promote MB bioadhesion, we anchored on the MB surface the arginine-glycine-aspartic tripeptide, i.e. RGD, and cysteine by reductive amination. The peptide molecule represents the structural motif of any proteins involved in a bioadhesion process. The binding of this peptide sequence to heterodimeric cell surface receptors called integrins is the process at the base of adhesion between the cells and extracellular matrix. The RGD motif is by far the most effective and the most employed peptide sequence used as "biomolecular glue", to stimulate cell adhesion on synthetic materials [10]. Coupling of MB surface with RGD was confirmed qualitatively by electrospray ionization mass spectrometry performed on a diluted suspension of RGD-functionalized MB5R. In Figure 3.4 the ESI-MASS spectrum of RGD decorated surface is shown. An RGD molecular peak at $m/z = 343$ followed by peaks relative to RGD $\times$ Na$^+$, RGD $\times$ 2Na$^+$ and RGD $\times$ 3Na$^+$, at $m/z = 365, 383$ and 403, respectively, are reported in Figure 3.4. The quantitative determination of RGD tethered on MBs performed by the microBCA colorimetric assay [3] yielded a value of 5×10^{-2} μmol RGD per mg of MB for all three types of MBs.

The design of MBs for the treatment of thrombosis and cardiovascular disease, requires a suitable conjugation strategy to the particle surface. Introduction of L-cysteine can be considered a "pool" of NO because of its potential role in nitric oxide storage and transfer in a range of physiological processes [4]. Ellman's test determination of the amount of L-cysteine yielded a value of 8×10^{-2} μmol of thiol per mg of MB. The L-cysteine functionalization allowed a chemical way to produce NO in the domain of the microballoons. This was achieved by cysteine nitrosation using NaNO$_2$ in HCl 0.1M. The corresponding nitrosothiol was obtained. The reaction was carried out at 37°C for 30 minutes. The functionalized MBs were washed three times with HCl

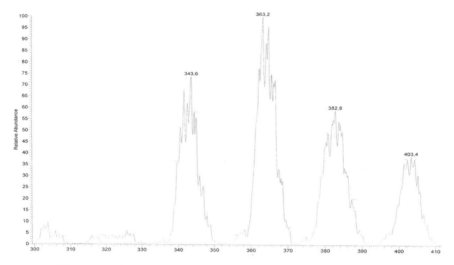

Fig. 3.4. ESI-MASS spectrum of RGD modified MB PVA shell

0.1M, twice with water and then suspended in PBS solution. At pII 7.4, NO is generated by the substrate and released for the proper specific application. This design can offer several advantages in the therapeutic administration of NO, ranging from an increased stability compared to NO donors in aqueous solutions, to the possibility to control the starting NO release compared to its release when NO gas is physical and unspecifically loaded on MBs.

The chemical route of MB nitrosation and NO production is reported in Scheme 2.

Due to the lability of NO, its detection in an aqueous medium is performed by using myoglobin as a quenching molecule, in the presence of dithionite to insure reducing conditions. The NO released by the MBs forms a complex with myoglobin which can be detected by means of electronic paramagnetic resonance, EPR. The spectrum of the complex, indicative of the six coordinate nitrosyl-heme complex with the characteristic $g_1 = 2.08$, $g_2 = 2.01$ and $g_3 = 1.98$ can be regarded as a fingerprint and is reported in Figure 3.5 [7].

O—CH$_2$-NH-CH-CH2-SH $\xrightarrow[\text{H}^+]{\text{NO}_2^-}$ O—CH$_2$-NH-CH-CH2-S-NO
    | |
    COOH   COOH

$\downarrow$ pH= 7.4

O—CH$_2$-NH-CH-CH2-S +
    |
    COOH

Scheme 2. Release of NO after L-cystein coupling to the microballoon surface

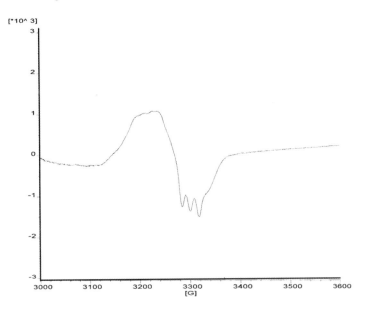

Fig. 3.5. NO-myoglobin EPR spectrum at 100 K

Cyclodextrins are often used in food and in pharma industries for their ability to form complexes with hydrophobic species [17]. Their inclusion properties can be suitably applicable for drug delivery strategies [5,16]. Moreover, the hydroxyl moiety exposed to the solvent offers an easy way for conjugation to the MB surface. We have decorated the MB surface with γ-cyclodextrin (γ-CD) because of its ability to complex doxorubicin (DOXO), as demonstrated by circular dichroism studies [1]. γ-CD was partially oxidized (one glucose unit per ring) with sodium metaperiodate and conjugated to the hydroxylic groups on the MB surface. The determination of the amount of cyclodextrin linked to the surface was accomplished by titrating spectrophotometrically the CD-derivatized MBs with phenolphthalein and monitoring the change in absorbance at 550 nm [11].

Results from this study was 4×10^{-8} mol γ-CD per mg of MB. Our future work will focus on the biological activity of DOXO-γ-CD complexes at the MB surface.

3.3.3 RGD functionalized PVA, RGD-PVA

The development of a general strategy for activating the surface of PVA based devices involves the efficient enabling of its conjugation with molecules of interest. As the main focus of this work is the surface modifications of PVA based MBs, several reactive routes were preliminarly explored on the flat surface of membranes prepared with the same chemistry used for MB fabrication.

We selected the coupling chemistry of nitrophenyl chloroformate, NPC, as a possible strategy for the conjugation of RGD to the polymer support [18]. As reported in Scheme 3, the hydroxyl groups of the PVA chains were reacted with NPC in 1-Methyl-2-pyrrolidinone, in the presence of pyridine and DMAP. After 24 hours at room temperature, the polymer was precipitated with ethanol. The white precipitate was collected and washed with ethanol until disappearance of the yellow color developed by the ethanol washings when poured into alkaline medium due to the presence of unreacted NPC.

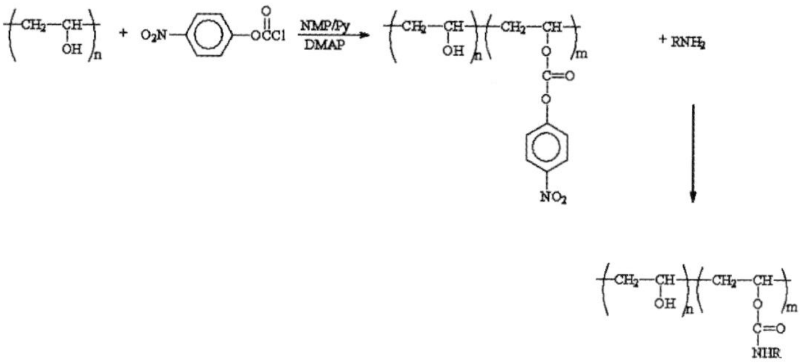

Scheme 3. PVA derivatization via nitrophenyl chloroformate, NPC, coupling chemistry

The p-nitrophenyl conjugated derivative was used as a useful intermediate for grafting RGD tripeptide onto the PVA chains. It is well known, that by reacting NPC functionalized substrates with molecules containing a primary amine in organic solvents and in the presence of a catalyst, the p-nitrophenolate is removed and a urethane group is formed. In order to obtain a PVA carrying the RGD tripeptide, we applied this reaction to PVA-NPC using DMSO as a solvent and pyridine and DMAP as catalysts. The reaction was carried out for 3 days at room temperature. To remove 4-nitrophenol and unreacted RGD, the reaction mixture was dialyzed against water at pH 11 and then against MilliQ water. The presence of coupled RGD was confirmed by circular dichroism as shown in Figure 3.6 where the spectra of RGD and PVA-RGD, via PVA(NPC) intermediate, in water are reported. Interestingly, an inversion of the ellipticity is revealed in the RGD coupled PVA aqueous solution. PVA-RGD was further characterized in solution by ^{1}H-NMR spectroscopy. The amount of coupled molecules to the PVA, determined by integration of RGD peak resonances, was equal to 2.5 % mol of RGD / PVA repeating unit (see Fig. 3.7).

3.3.4 Synthesis of RGD-PVA hydrogels

Telechelic PVA was obtained by the selective oxidation of the head-to-head sequence of PVA chains by metaperiodate. For the telechelic hydrogels, the

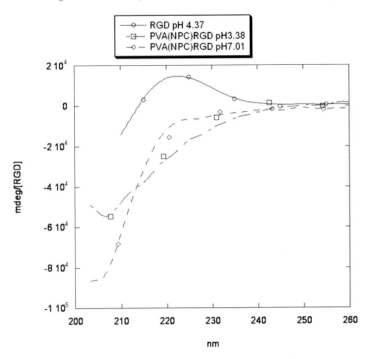

Fig. 3.6. Circular dichroism spectra of aqueous solution of RGD at pH 7 and of PVA-RGD, obtained from the PVA(NPC) intermediate, at pH 3.38 and 7.01

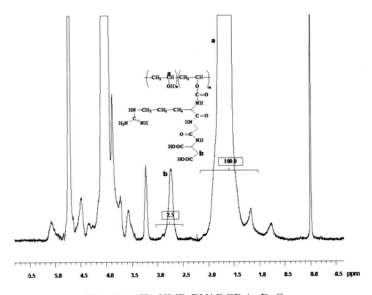

Fig. 3.7. H1-NMR PVARGD in D_2O

membrane formation was obtained by a cross-linking reaction between reactive aldehydic end groups of PVA chains and PVA hydroxyl groups. The PVA-RGD based network formation was carried out by a two steps-one pot procedure:

- a telechelic RGD-PVA macromer carrying aldehydic groups as chain ends was obtained by the selective oxidation of the head-to-head sequence of RGD-PVA chains by metaperiodate, followed by;
- a cross-linking reaction performed by an acetalization reaction of telechelic PVA with the telechelic PVA-RGD macromer.

The estimated average surface density of RGD on hydrogel membranes was found to be equal to 70 fmol of RGD molecules/cm^2.

3.4 Conclusions

In this chapter we explored different strategies for the surface modification of MBs. The main goal was to assess the effective possibility to functionalize PVA MB surface with biologically relevant molecules and to evaluate the real chemical versatility of the MB surface, one of the main points outlined in Table 3.1.

PVA offers several conjugation options. However, surface MB modification imposes limitations in the choice of reaction medium to use for the surface modification reactions. Organic good solvents for PVA are not allowed as they contribute to the partial dissolution of the cross-linked polymer chains. In general, organic solvents having a smaller surface tension than water, destabilize the microballoons. Acetalization and reductive amination reactions provide an effective route to couple in an aqueous environment hydroxyl and aldehyde groups present on the hydrophillic MB surface with molecules bearing aldehyde and amino groups. In the chapter "Polymer based biointerfaces: a case study on devices for theranostics and tissue engineering." in vitro studies on biocompatibility of some of the modifed MBs are reported.

For its quantitative character and for the possibility to use water as a reaction medium, click chemistry offers, in principle, new and important perspectives in the surface derivatization of PVA microballoon surface. Exploratory studies are underway in our laboratory.

References

1. AI-Omar A, Abdou S, De Robertis L, Marsura A and Finance C (1999) Complexation study and anticellular activity enhancement by doxorubicin-cyclodextrin complexes on a multidrug-resistant adenocarcinoma cell line. Bioorg Med Chem Lett 9:1115–1120
2. Ando I et al. (2000) NMR spectroscopy in polymer science. In Tanaka T (ed.) Experimental Methods in Polymer Science, Academy Press, San Diego

3. Banerjee P, Irvine DJ, Mayes AM and Griffith LG (2000) Polymer latexes for cell-resistant and cell-interactive surfaces. J Biomed Mater Res 50:331–339
4. Bohl KS and West JL (2000) Nitric oxide-generating polymers reduce platelet adhesion and smooth muscle cell proliferation. Biomaterials 21:2273–2278
5. Brewster ME and Loftsson T (2007) Cyclodextrins as pharmaceuticals solubilizers. Adv Drug Delivery Rev 59:645–666
6. Cavalieri F, El Hamassi A, Chiessi E and Paradossi G (2005) Stable polymeric microballoons as multifunctional device for biomedical uses: synthesis and characterization. Langmuir 21:8758–8764
7. Cavalieri F, Finelli I, Tortora M, Mozetic P, Chiessi E, Polizio F, Brismar T and Paradossi G (2008) Polymer microbubbles as diagnostic and therapeutic gas delivery device. Chem Mater 20:3254–3258
8. Chiessi E, Cavalieri F and Paradossi G (2007) Water and polymer dynamics in chemically cross-linked hydrogels of poly(vinyl alcohol): a molecular dynamics simulation study. J Phys Chem B 111:2820–2827
9. Gombotz WR, Guanghui W, Horbett TA and Hoffman A (1991) Protein adsorption to poly(ethylene oxide) surfaces. J Biomed Biomater Res 25:1547–1562
10. Hersel U, Dahmen C and Kessler H (2003) RGD modifed polymers: biomaterials for stimulated cell adhesion and beyond. Biomaterials 24:4385–4415
11. Holm R, Hartvig RA, Nicolajsen HV, Westh P and Ostergaard J (2008) Characterization of the complexation of tauro- and glyco-conjugated bile salts with γ-cyclodextrin and 2-hydroxypropyl-γ-cyclodextrin using affinity capillary electrophoresis. J Inclusion Phenom Macrocyclic Chem 61:161–169
12. Kaufmann BA and Lidner JR (2007) Molecular imaging with targeted contrast ultrasound. Curr Opin Biotecnol 18:11–16
13. Klibanov AL (2005) Ligand-carrying gas-filled microbubbles: ultrasound contrast agents do targeted molecular imaging. Bioconjugate Chem 16:9–17
14. Klibanov AL (2007) Ultrasound molecular imaging with targeted microbubble contrast agents. J Nucl Cardiol 14:876–884
15. Liu Y, Miyoshi H and Nakamura M (2006) Encapsulated ultrasound microbubbles: therapeutic application in drug/gene delivery. J Controlled Release 114:89–99
16. Stella VJ, Rao VM, Zannou EA and Zia V (1999) Mechanism of drug release from cyclodextrin complexes. Adv Drug Delivery Rev 36:3–16
17. Szente L and Szetjli J (2004) Cyclodextrins as food ingredients. Trends Food Sci Technol 15:137–142
18. Taniguchi H, Akiyoshi K and Sunamoto J (1999) Self-aggregate nanoparticles of cholesteryl and galacoside groups-substituted pullulan and their specific binding to galactose specific lectin, RCA120 200:1554–1560
19. Thiele L, Diederichs JE, Reszka R, Merkle HP and Walter E (2003) Competitive adsorption of serum proteins at microparticles affects phagocytosis by dendritic cells. Biomaterials 24:1409–1418
20. Unger EC, Porter T, Culp W, Labell R, Matsunaga T and Zutshi R (2004) Therapeutic application of lipid-coated micro bubbles. Adv Drug Delivery Rev 56:1291–1314

Chapter 4

Lipidic Microbubble Targeting of Surface Proteins Using an in Vitro System

Ines Zerbe and Nenad Gajovic-Eichelmann

Abstract. Lipidic microbubble contrast media containing targeting molecules, e.g. antibodies specific for some tumor-related protein, in their outer shell, are a promising strategy to improve the contrast of tumor lesions in ultrasound imaging. The preparation and characterization of such targeted lipidic microbubbles is a difficult task mainly because of their short lifespan and fragile structure as well as their continuous "floating to the surface" due to buoyancy. Moreover, few cancer-related surface proteins are available in pure form. Classical immunochemical tests, like ELISA, cannot be employed for the analysis of targeted lipidic microbubbles. Instead, cancer cell lines can be used for in vitro experiments. Two different types of in vitro experiments are presented, both addressing the tissue factor (TF) as the tumor-related protein and using lipidic soft-shell microbubbles with attached monoclonal anti-TF antibody as the contrast media. Flow cytometry was used to analyze the interaction between targeting microbubbles and U-937 cancer cells growing in suspension. Perfusion in a microfluidic channel was used to study the interaction between microbubbles and the surface adherent cell line U87-MG.

4.1 Introduction

Direct visualization of cancer tissue by cancer-specific, molecularly targeted contrast media is the ultimate goal of ultrasound imaging [3]. Due to the 1–7 μm size of typical contrast microbubbles, they are restricted to the vasculature. As a consequence, only vascular tumor markers can be addressed. Besides the different vascular epithelial growth factor receptors (VEGFR) [1], the tissue factor (TF) has recently been found of interest as a potential vascular marker. TF is a 47 kDa membrane protein expressed only in the basal lamina but not in the endothelium of blood vessels in healthy tissue. Upon damage of the endothelium and binding of blood-derived Factor VII to TF, the blood clotting cascade is initiated. TF also has been reported to be expressed in tumor vessels in prostate cancer [2]. Two commercially available human cancer cell lines, which are known to express TF constitutively are U-937, a non-adherent lymphoma cell line, growing in suspension [4]. The astrocy-

Paradossi, G., Pellegretti, P., Trucco, A. (Eds.)
Ultrasound contrast agents. Targeting and processing methods for theranostics
© Springer-Verlag Italia, 2010

toma cell line U87-MG is an adherent cell line, growing exclusively on surfaces. Monoclonal antibodies binding to the extracellular domain of TF have been described and are commercially available. Therefore, TF is an interesting model protein for in vitro studies of molecularly targeted contrast media. Such studies are needed in order to optimize the microbubble formulation, to investigate the binding efficiency and binding specificity. Only after in vitro optimization of the targeted microbubbles, is an in vivo experiment addressing a real tumor justified.

The results obtained from investigating binding of targeted microbubbles to suspension cells (U-937) and to adherent cells (U87-MG) are of different quality: Binding of free floating microbubbles to free floating cells (U-937) in a large volume of a slightly mixed aqueous solution is an idealized situation, not very similar to the in vivo scheme. Due to the absence of surfaces and boundaries, allowing both bubbles and cells to move relative to each other, eventually a binding equilibrium will be reached. This equilibrium should, as a first hypothesis, reflect the affinity of the targeting antibody to its target protein, the density of antibodies on the microbubbles, the density of target proteins on the cells, and the concentrations of bubbles and cells. If the concentrations of microbubbles and cells in an experiment can be measured and reproduced within different experiments, the difficult-to-measure intrinsic molecular properties of bubbles (and cells) should govern the binding behavior and the most efficient microbubble preparations can be detected by comparing such equilibrium binding data. Flow cytometry is a suitable detection method for this type of experiment: Using the principle of hydrodynamic focusing in a sheath flow of saline, different data as light scattering properties and fluorescence can be generated for each single cell which provides information about cell size, and destruction and interaction of the cell with fluorescence-labelled microbubbles. Counting of targeting microbubbles by exploiting light scattering is in principle also possible, but the microbubbles are much easier to quantify after fluorescent labeling.

Binding of microbubbles to surface adhered cells (U87-MG) by perfusion in a single, rectangular microfluidic channel resembles much more the situation in blood vessels: Due to the buoyancy force all floating microbubbles will eventually reach the top of the channel, thereby pushing the immobile cells lining the top face of the microfluidic channel. In effect, a much closer contact between individual bubbles and cells will arise, resulting in a higher binding probability. Previous experiments have shown that in such condition the number of bound microbubbles per square area is linear with time, until the surface is saturated, i.e. a significant portion is covered with bubbles. In order to compare different microbubble preparations, one needs to control the flow rate, the concentration and size distribution of microbubbles, the surface coverage of the cell layer ("confluency") and the interaction time. An optical bright-field microscope equipped with a camera, attached to a PC with image processing software, is a suitable detector for such perfusion experiments. The perfusion chamber must be transparent and have a well defined flow geometry. In addition, phase contrast microscopy may be a helpful imaging modality to visualize the cell layer. Even with a standard microscope it is possible to count bound microbubbles, but it is almost impossible to count all floating bubbles during a 10 min experiment. Because of that, a Coulter-counter is also needed to detect the exact microbubble concentration (and size distribution) beforehand.

The main drawbacks of the perfusion experiment are a high unspecific binding of microbubbles, mainly caused by physical entrapment in gaps between cells and the difficulty to prepare highly reproducible cell layers. Therefore, each microbubble targeting experiment is accompanied by one experiment using isotype control microbubbles, i.e. microbubbles equipped with a mixture of antibodies similar to the targeting antibody, but without the antigen specificity required for targeting. By calculating the ratio between surface bound targeting and isotype microbubbles (RTI) under the same experimental conditions, we obtain a robust figure of merit for the "real" molecular targeting efficiency.

4.2 Material and methods

4.2.1 Antibodies and reagents

Streptavidin coated lipidic microbubbles (BG 6128, diameter around 1.5–5 µm) as described in the chapter "Use of lipidic ultrasound contrast agent for molecular imaging" were kindly provided by Bracco Research S.A. (Geneve, Switzerland). Purchased antibodies were mouse anti-human TF mAb clone TF9-10H10 (Merck Biosciences, Darmstadt, Germany), anti-human TF mAb clone ad4509 (American Diagnostica, USA), FITC- goat anti-mouse IgG (Rockland, Gilbertsville, USA). Biotinylation of TF9-10H10 antibodies was performed with ChromaLink Biotin 354S biotinylation kit (SoluLink, San Diego, USA) to an average coupling rate of 2–3 Biotins per molecule. Quantification was done by spectrophotometry with NanoDrop UV/VIS spectral photometer (Peqlab, Erlangen, Germany).

4.2.2 Cell culture of U-937 for flow cytometry testing

U-937 human lymphoma cell line (DSMZ, Braunschweig, Germany) was grown in a humidified incubator at 5% CO_2 and $37°C$ in RPMI 1640 cell culture medium (Biochrom, Berlin, Germany) supplemented with 10% fetal calf serum (Biochrom, Berlin, Germany). Subcultivation was carried out every 3 days. After harvesting, cells were washed 2 times in PBS and afterwards diluted in PBS containing 0.5% BSA.

4.2.3 TF receptor binding assay

Streptavidin coated microbubbles were reconstituted from lyophilisate with 0.9% NaCl solution to a final concentration of around 10^9 bubbles ml^{-1}.

10^9 bubbles were incubated with 15 µg of biotinylated TF9-10H10 antibody for 10 min at RT in a tube rotator. Aliquots of 2.5×10^5 U-937 cells in 1 mL of phosphate-buffered saline supplemented with 0.5 % BSA were incubated with 5×10^6, 1×10^7, 5×10^7 or 1×10^8 antibody coated microbubbles or plain microbubbles, respectively, and stained with 50 µg/mL FITC labelled secondary antibody for 40 min at 37°C, then centrifuged (180 g, 10 min), washed twice, resuspended in isotonic sodium chloride solution and measured immediately by flow cytometry.

4.2.4 Flow cytometry

For cytometry a FC500 flow cytometer (Beckman-Coulter, Krefeld, Germany) with 488 nm air cooled argon laser as a light source was utilized. Isotonic Sodium Chloride Solution (0.9 % (w/v) NaCl in ultrapure water) was used as a sheath fluid and for dilution. The standard cell number collected was 30,000 cells per run. Gating was carried out on FS and SS parameters and data processing was done with CXP analysis software.

4.2.5 Cell culture of U87-MG for perfusion testing

U87-MG human astrocytoma cell line (ECACC, UK) was cultivated on disposable polymer slides with 6 integrated microchannels (µ-Slide VI, tissue culture treated & sterile, Ibidi, Germany; microchannel dimensions are 3.8 mm width, 0.4 mm height, 17 mm length, 0.6 cm^2 growth area) in a humidified incubator at 5 % CO$_2$ and 37°C in MEM cell culture medium (Biochrom, Germany) supplemented with 1 % HEPES (Applichem, Germany), 1 % non-essential amino acids (Biochrom, Germany), 1 % Sodium pyruvate (Applichem, Germany) and 10 % fetal calf serum (Biochrom, Germany). 10^4 cells, dispensed in 30 µl medium, were seeded per microchannel. The Ibidi µ-Slide was then turned upside down and cultivated for 2 hours, to allow cell attachment to the top of the flow channel. Finally the µ-Slide was turned again, and the reservoirs were filled with fresh medium. Cultivation was continued for 2–3 days with cells growing on top of the microchannel, until 60–80 % confluency was obtained. No fresh medium was supplemented during this time.

Expression of TF by cultivated U87-MG cells was tested by TF Immunostaining according to the following protocol: Cells were pre-washed 3 times with 100 µl each of 37°C PBS. 50 µL of anti-TF mAb (at 6 µg mL^{-1}, in 37°C PBS) was added to the microchannel, and incubated for 30 min (at 37°C). Cells were washed 6 times with 100 µl of pre-warmed PBS, before FITC-goat anti-mouse IgG (at 12.5 µg mL^{-1}, in 37°C PBS) was added. After a 30 min incubation, and a final 6 washes as before, cells were imaged under the fluorescence microscope (Olympus iX50, FITC-channel, 10 × magnification, Camera: Olympus 2 Mpix). As the negative control, cells treated as before with an isotype antibody mix instead of anti-TF mAb, were used.

4.2.6 Maximum shear stress determination

U87-MG cells growing at the top and bottom of an µ-Slide VI perfusion cell were subjected to a hydrodynamic flow of 0.9 % NaCl solution at flow rates ranging from 19 µl min^{-1} up to 614 µl min^{-1}, produced with a peristaltic pump (DU-520, Watson-Marlow, 0.38 mm iD tubing) connected to the outlet of the microchannel by a Luer connector. The feed reservoir was filled with 0.1 ml 0.9 % NaCl solution, taking care not to introduce any air bubbles into the channel. The cells were inspected visually by phase contrast microscopy during the perfusion experiment, starting with the lowest flow rate and doubling the flow rate every 3 minutes. The reservoir was replenished several times without stopping the flow. The critical shear rate was detected by noticing the beginning rounding off of cells, preceding their detachment from the surface.

4.2.7 Perfusion experiment with U87-MG

Streptavidin coated microbubbles were reconstituted from lyophilisate with 0.9 % NaCl solution to a final concentration of around 10^9 bubbles mL^{-1} just before the perfusion experiment. Targeted microbubbles were incubated with 15 µg/mL biotinylated TF9-10H10 antibody for 10 min at RT in a tube rotator. Isotype control bubbles were incubated accordingly with 15 µg/mL of biotinylated isotype control antibody mix. The microbubble concentration and size distribution of both preparations was determined with a Coulter-Counter equipped with a 100 µm capillary (Casy 1, Germany), typically at a working dilution of 1:10000.

A µ-Slide perfusion cell with cultivated U87-MG cells was connected to the perfusion pump as described before. A scheme of the setup is depicted in Figure 4.1. The feed reservoir was filled with 0.1 mL of 0.9 % NaCl solution and flow was applied until the feed was almost empty. Microbubble solution at a concentration of around 100,000 bubbles mL^{-1} was added to the reservoir with a 0.5 mL microsyringe. The perfusion started by applying a flow rate

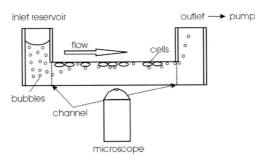

Fig. 4.1. Schematic setup of the microbubble binding experiment in a µ-Slide VI perfusion chamber

of 76 µL min^{-1} (shear rate = 12.5 sec^{-1}, shear stress = 0.125 dyn cm^{-2}). Care was taken not to introduce any air-bubbles during the experiment, by replenishing the reservoir during perfusion. Direct addition of microbubble suspension to the reservoir was preferred to a supply by a connected tubing, since microbubbles tended to accumulate in bends and kinks of the tubing, and a constant bubble supply during perfusion was more difficult to achieve. The flow was stopped after 10 minutes.

The detection was accomplished in real-time by image aquisition of bright-field microscopy images at a rate of 1 image sec^{-1} (by Leica DMR phase contrast & brightfield inverse microscope, with 10 × magnifying phaco objective, Leitz, Germany. Camera was a DVT Legend 544, DVT, Canada; 1024*1280 pixel, field of view 0.92 mm × 1.1 mm). An exposure time of only 0.3 msec was sufficient to obtain sharp, high-contrast images. An additional phase contrast image of the cell layer was taken before and after perfusion (10 × magnification, 3 msec exposure).

4.2.8 Image analysis

High contrast images of bound microbubbles were obtained by subtracting of an image taken before perfusion, from the image taken after perfusion. The subtraction increased the contrast of bubbles (visualized as black dots) and eliminated most of the distracting details of cell images. The freeware ImageJ 1.38X from the NIH (http://rsb.info.nih.gov/ij) was used for this purpose (operation: image processing//subtraction). Difference images were output as memory-friendly 32-bit greyscale Jpg. Counting of bound bubbles was done by manual counting on the screen.

4.3 Results

4.3.1 Flow cytometry investigation of targeted microbubbles-cell interactions

To investigate the interaction of U-937, a cell line growing in suspension and expressing the specific molecular marker TF, with an ultrasound contrast agent targeted to TF, we modified streptavidin coated microbubbles with biotinylated monoclonal antibodies recognizing this protein. Because the attachment of microbubbles to the cell surface can not be recognized by size measurement via forward light scatter due to the non-compact nature of gas-filled microbubbles, the bubbles were labelled with a fluorescent secondary antibody to get a clear binding signal. The parameter of interest is the median fluorescence intensity (MFI) of the cell population. As negative controls we used microbubbles with no functional surface viz. not pre-incubated with TF-antibody (Fig. 4.2a) and functional microbubbles without fluorescence-labelled secondary antibody (data not shown). After incubation

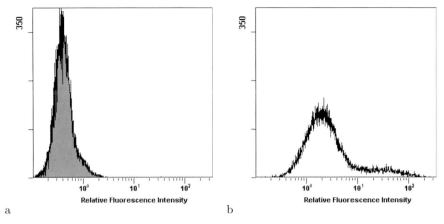

a   b

Fig. 4.2. Flow cytometry of U-937 cells after incubation with TF targeted microbubbles: Fluorescence intensity distribution of a population of 30,000 U-937 cells incubated with (**a**) non-targeted, but fluorescence labeled and (**b**) targeted and fluorescence labeled microbubbles. Targeted microbubbles show a 5-fold increase in the average fluorescence intensity (median fluorescence 2.16) relative to the non-targeted bubbles (median fluorescence 0.38)

with targeted microbubbles we monitored an up to 5-fold increase of MFI relative to the unspecific control (Fig. 4.2b).

To investigate the concentration dependence of the binding of targeted microbubbles to free floating cells, we incubated U-937 cells with an increasing excess from 20-fold up to 400-fold of targeted microbubbles as described above. The unspecific binding was estimated by the use of non-functional microbubbles instead of targeted ones. Plotted is the ΔMFI (the median fluorescence intensity of the cell population subtracting the median fluorescence intensity of the unspecific control) against the number of microbubbles per cell incubated (Fig. 4.3). The mean fluorescence per cell is linearly increasing with the excess of targeted bubbles indicating that there is no saturation of binding in this range.

4.3.2 Cell culture of U87-MG for tissue factor targeting in a perfused microchannel

Human astrocytoma cells U87-MG were cultivated to 60–80 % confluence at the top of rectangular microchannels of disposable polymer perfusion chambers (Ibidi μ-Slide VI). Expression of the target protein TF was verified by immunostaining and fluorescence microscopy. Figure 4.4 shows the images obtained with the FITC fluorescence channel (excitation wavelength 495 nm, emission wavelength 515 nm, 0.2 sec exposure time) for cells stained with the anti-TF mAb TF9-10H10 (Fig. 4.3a) and for cells stained with isotype control antibody mix (Fig. 4.3b). Anti-TF treated cells exhibited a strong surface

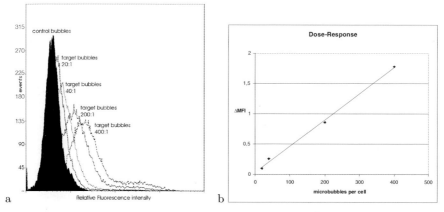

Fig. 4.3. Dose-response plot for flow cytometry of 2.5×10^5 U-937 cells after incubation with an increasing excess of TF targeted microbubbles. (**a**) The histogram plots show an increasing average fluorescence level of the cells on the x-scale (**b**) Dose-response curve from the histogram plots. The depicted median fluorescence values (ΔMFI) are corrected for unspecific binding, which was assessed by previous experiments with non-targeted, streptavidin-coated microbubbles. No saturation of binding occurs

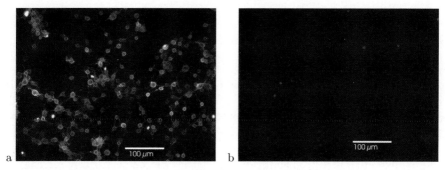

Fig. 4.4. Fluorescence microscopy images of U87-MG cells which have been (**a**) treated with anti-TF mAb and (**b**) treated with isotype antibody mix after immunostaining with a FITC-labeled goat anti-mouse secondary antibody ($10 \times$ magnification, field-of-view 0.6 mm $\times 0.4$ mm, integration time 0.2 sec)

fluorescence, which could be visually observed, indicating strong expression of TF at the surface of U87-MG cells. In contrast, the isotype control treated cells show almost no fluorescence. The upside-down mode of cell cultivation obviously did not affect the cells.

4.3.3 Shear stress tolerance of U87-MG

During the perfusion experiments with 0.9% NaCl solution at room temperature, U87-MG cells are facing multiple stress factors. Temperature drops

Table 4.1. Shear stress tolerance of U87-MG cells cultivated at the top of the microchannel of μ-Slide VI

flow rate ($\mu L\ min^{-1}$)	shear rate (sec^{-1})	shear stress ($dyn\ cm^{-2}$)	visual effect on cells
19	3.1	0.031	none
38	6.2	0.062	none
78	12.5	0.125	none
153	25.2	0.251	none
307	50.3	0.503	none
614	100.6	1.006	cells rounded

from 37°C to around 20°C, the medium is deprived of any nutrients and the flow exhibits shear stress on adherent cells. Since typical perfusion experiments lasted only between 10 and 20 min, starvation or cold adaptation were excluded. Shear stress, however, can have an immediate effect on cells (detaching, damage), therefore perfusion has to be performed under conditions far from the critical shear stress limit. Table 4.1 summarizes the results. Shear rate and shear stress were calculated using the formulas supplied by Ibidi for μ-Slide VI channel geometry. Only at the upper limit of 614 μL min^{-1} was an effect on cells observed within 3 min of perfusion. For the routine perfusion experiments, a flow rate of 80 μL min^{-1} was selected, far from the critical shear stress limit.

4.3.4 Tissue factor targeting with microbubbles in a perfused microchannel

In the typical experiment, each cell-grown microchannel was perfused only once with targeted or isotype microbubble suspension for 10 min at a flow rate of 80 μl min^{-1} (shear rate 100 sec^{-1}, shear stress 0.125 dyn cm^{-2}) to avoid physiological changes and cell damage during the experiments. In an alteration to this scheme, some microchannels were first perfused with targeted microbubbles for 10 min, and then with isotype control microbubbles (or vice versa). The purpose was to discriminate specific and unspecific binding events to exactly the same cell layer. Microbubbles bound to the surface were most easily counted from the difference images, as described above. Targeted and isotype control microbubble experiments were evaluated separately. The ratio of bound targeting bubbles and isotype bubbles (under the same experimental conditions) was calculated as a figure of merit for the binding specificity. Figure 4.5 illustrates the time dependence of target bubbles (and control bubbles) binding to the cell layer. The preferred binding of targeted bubbles is obvious. The graphs are not strictly monotonic, because bound bubbles can be lost during the timecourse of the experiment either by detach-

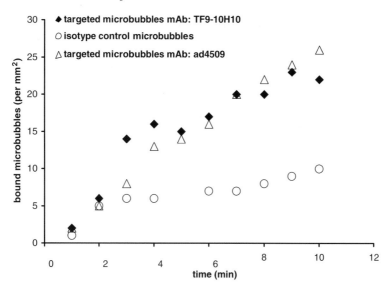

Fig. 4.5. Binding of targeted and control microbubbles to adherent cells U87-MG: Time dependence (targeting antibody: TF9-10H10 and ad4509, bubble concentration around 10^5 mL $^{-1}$. Flow rate 80 μL min $^{-1}$). Binding events increase with time in a linear fashion

ment or bursting. Figure 4.6 shows a difference image after binding of targeted bubbles (Fig. 4.6a) and after subseqent binding of control bubbles to the same microchannel (Fig. 4.6b). Table 4.2 summarizes a set of multiple experiments obtained with two different monoclonal antibodies against TF, at two different microbubble concentrations and at different flow rates. In all instances the

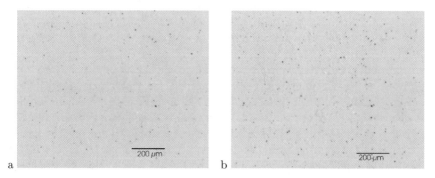

Fig. 4.6. U87-MG cells at the surface of the microchannel after (**a**) 10 min perfusion with targeted and (**b**) a second 10 min perfusion with control microbubbles (on top of targeted bubbles) at 80 μL min^{-1}. Concentration was around 10^6 bubbles mL^{-1}. Bubbles appear as clear black dots in the difference images, while the contrast of cells is highly reduced. White spots are footprints of initially present bubbles, which were lost during the experiment

Table 4.2. Summary: Tissue factor targeting with microbubbles in a perfused microchannel

antibody	concentration (bubbles mL^{-1})	flow rate ($\mu L\ min^{-1}$)	targeting bubbles bound per mm^2	control bubbles bound per mm^2	targeting/ isotype ratio
ad4509	10^6	19	221	183	1.21
ad4509	10^6	19	261	183	1.43
ad4509	10^6	80	349	187	1.87
ad4509	10^5	80	26	11	2.36
ad4509	10^5	80	37	11	3.36
TF9-10H10	10^6	80	254	187	1.35
TF9-10H10	10^6	80	250	187	1.34
TF9-10H10	10^6	80	224	187	1.20
TF9-10H10	10^5	80	28	11	2.55
TF9-10H10	10^5	80	22	11	2.0

binding of targeted microbubbles was preferred over control bubbles. The targeting to isotype ratio changes from a minimum of 1.2 up to 3. From these data it appears that the specificity of binding increases with increasing flow rate, and with decreasing microbubble concentration. The two different monoclonal antibodies under test achieved the same level of specificity.

4.4 Conclusions

Two different in vitro approaches to characterize ultrasound contrast agents, targeted against the tissue factor receptor protein, were investigated. Targeting antibodies bearing a biotin group were bound to streptavidin-coated microbubbles. In this way, variation of the antibody was easily achieved. Flow cytometry was used to study the binding equilibrium between targeted and fluorescence labeled microbubbles and suspended cells, using the U-937 human lymphoma cell line. Large numbers of cells ($> 10^5$) and binding events can be easily counted, to minimize the statistical error. Binding of microbubbles to cells was most easily tracked by the accumulated fluorescence, which increased up to 5-fold. The implicit need to fluorescence-label the targeted contrast media and the impossibility to extract information about bubble size and shape from flow cytometry data are, however, disadvantages. Care must also be taken when interpreting flow cytometry data from adherent cell lines. The initial trypsin treatment may have an impact on their physiological state. Therefore, flow cytometry can be considered a useful and quick tool to optimize targeting microbubbles (e.g. with respect to ligand type, ligand density, etc.) when the target is expressed by a cell line growing in suspension.

Targeting of surface proteins expressed by adherent cells, like U87-MG, was accomplished in a perfused microfluidic channel using only standard equipment present in many laboratories, and a free software. Disposable cell culture and perfusion slides with integrated microfluidic channels and sample reservoirs (μ-Slide VI by Ibidi), connected to a simple peristaltic pump, proved to be a suitable tool for the in vitro study of targeted contrast media. Due to the flat geometry and high transparency of these polymer perfusion slides, a standard brightfield microscope (with a camera) was sufficient to obtain high contrast images of floating and stagnant microbubbles without fluorescent labeling. Simple image processing by subtraction (e.g. using the freeware ImageJ) produced high-contrast greyscale images of microbubbles, which could be counted by visual inspection. Several preparations of targeted microbubbles were compared using this system. The targeting efficiency and specificity was assessed by the target-to-isotype control ratio, i.e. the ratio of bound targeted bubbles normalized by the number of bound isotype control bubbles. This dimensionless figure reflects the amount of specific binding (i.e. "targeting") of targeted US contrast media to a cell layer. Values between 1.2 (low specificity of binding) and 3 (high specificity of binding) were achieved. The highest specificity was obtained with a high flow rate and a low concentration of bubbles (10^5 mL^{-1}) during perfusion. No differences were detected on changing the targeting antibody, probably because both monoclonal antibodies under study were high affinity binders. The in vitro characterization of targeted contrast media in a perfused microchannel proved to be an inexpensive yet sensitive and reproducible method to detect subtle changes in targeted ultrasound contrast media.

Acknowledgements

Soft-shell microbubbles with streptavidin coating ("target-ready microbubbles") were kindly donated by Bracco Research SA.

This project was funded by the EC in the FP6 programme, priority 2 IST & 3 NMP, under the project no. NMP4-CT-2005-016382.

References

1. Alitalo K and Carmeliet P (2002) Molecular mechanisms of lymphangiogenesis in health and disease. Cancer Cell 1:219–227
2. Hu Z and Garen A (2001) Targeting tissue factor on tumor vascular endothelial cells and tumor cells for immunotherapy in mouse models of prostatic cancer. PNAS 98(21):12180–12185
3. Klibanov AL (2007) Ultrasound molecular imaging with targeted microbubble contrast agents. J Nuclear Cardiology 14(6):876–884
4. Lindmark E, Tenno T and Siegbahn A (2002) Role of Platelet P-Selectin and CD40 Ligand in the Induction of Monocytic Tissue Factor Expression. Arterioscler Thromb Vasc Biol 20:2322–2328

Chapter 5

Protein Interactions with Microballoons: Consequences for Biocompatibility and Application as Contrast Agents

Johannes Stigler, Martin Lundqvist, Tommy Cedervall, Kenneth Dawson, and Iseult Lynch

Abstract. The role of proteins as the mediators of the interaction between engineered materials (biomaterials) and living systems has long been appreciated, but the subtleties and complexities introduced by changing surface curvature are only beginning to be understood. Thus, in implant devices, where the biomaterial is presented as a flat surface, a very limited range of proteins bind to the material, these being mainly albumin and fibrinogen. However, as the surface curvature increases (as we move towards micro and nano scale particles) novel effects are observed, and the materials begin to bind rarer specialized proteins with very high affinity, which has significant consequences for their biocompatibility and for their impacts on the biological system with which they interact. In the present work we present some findings from the EU project SIGHT in which the interactions of polymeric microballoons (gas-filled polymer-shelled devices that are being developed as contrast agents for theranostic applications) with plasma proteins are investigated, and the potential consequences for long term biocompatibility are discussed.

5.1 Introduction

Microballoons are gas filled polymeric shells which are being intensively researched as an alternative to the more conventional lipid-shelled microbubbles for use as a contrast agent in the detection (and potentially treatment) of disease via medical ultrasound diagnostics and therapeutics [15]. A key advantage of polymer-shelled microballoons over lipid microbubbles is that they retain their gas core over significantly longer periods, making them more robust and versatile. Additionally, the relatively high proportion of reactive aldehydic groups that remain at the surface following formulation of the microballoons makes them very suitable for modification with targeting ligands or therapeutic species, [3] opening up a route to multi-functional (theranostic) devices.

Ongoing development and refinement of the microballoons is being undertaken by the SIGHT consortium, who are iteratively studying the acoustic, mechanical and biological properties of the microballoons, in order to opti-

Paradossi, G., Pellegretti, P., Trucco, A. (Eds.)
Ultrasound contrast agents. Targeting and processing methods for theranostics
© Springer-Verlag Italia, 2010

mize all properties whilst ensuring that the long-term biocompatibility of the microballoons is maintained. Approaches to determining biocompatibility are varied, from proliferation studies as a function of time or assessment of DNA damage, [7] to detailed assessments of the interaction of the microballoons with serum/plasma proteins, as application via the bloodstream is the most likely route of administration for ultrasound contrast agents. It is well known from the medical device industry, that surfaces introduced into the body are immediately coated with a layer of proteins, and that it is this interaction that determines the success (or otherwise) of the implant [1, 18]. More recently, problems have emerged with polymer-coated metal stents which elute drugs (so-called Drug Eluting Stents) over significant time periods (up to two years) as a result of late stage thrombosis [14, 17]. This occurs because of longer term bio-incompatibility issues that are not detected by conventional toxicity testing, but may be detectable by changes in the adsorbed protein layer or other more advanced approaches to determining biocompatibility.

A hypothesis that has been receiving significant attention over the last couple of years is that the interaction between micro- and nano-scale materials with living systems is mediated by the dynamic layer of proteins which adhere to the materials immediately upon contact with biological fluids [5, 9, 12]. Clearly as the size of the particles decreases, the surface curvature increases, and when the particles are small enough, they can even be on the same scale as the proteins that they are interacting with, as shown in Figure 5.1. The

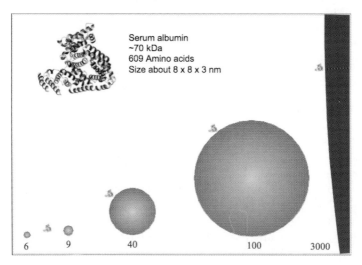

Serum albumin
~70 kDa
609 Amino acids
Size about 8 x 8 x 3 nm

6   9   40   100   3000

Fig. 5.1. Size comparison of several types of particles in the micro and nano range with serum albumin. The size is given in nanometers (3000 nm = 3 microns). In terms of protein binding, a size dependence is expected for the smaller particles where the particle is approaching the size of the proteins. Bigger particles appear flat to the proteins. Size dependence is therefore not expected in this regime, except where surface roughness or porosity is present

microballoons studied here are approximately 3 micron in diameter, and thus may resemble flat surfaces with low curvature. However, as a result of their polymeric and cross-linked nature, they very likely have different densities of polymer at the surface leading to surface roughness which can impact on their interactions with proteins.

As a consequence of their likely application in blood as an ultrasound with contrast agent, the first biological fluid that the microballoons will have contact is blood. As a model for blood, we use human plasma, which is stabilized with EDTA to arrest the coagulation cascade, which changes the nature of the system, and removes many of the key proteins. The interaction of the PVA-shelled microballoons with whole plasma and representative plasma proteins has been investigated in this work. In particular, we were interested to identify the major proteins in the microballoon protein corona, and to begin to understand the potential biological impacts of the microballoons on the basis of their associated proteins, which gives them their biological identity.

Plasma as a biological fluid is highly complex, containing over 3700 proteins [13]. There is a very large natural variation in protein composition and expression, even between healthy individuals. For example, the protein composition of blood (plasma) has been shown to vary significantly between individuals, with many of the proteins that are considered the wild-type not being the one present in the majority of individuals [13]. However, just a few proteins constitute the majority of plasma, being highly abundant: 60 % of plasma is albumin, which is a major contributor to the osmotic pressure of plasma, and assists in the transport of lipids and steroid hormones. Globulins make up 35 % of plasma and are involved in the transport of ions, hormones and lipids assisting in immune function. Fibrinogen comprises 4 % and is essential in the clotting of blood and can be converted into insoluble fibrin. Regulatory proteins which make up less than 1 % of plasma proteins are proteins such as enzymes, proenzymes and hormones [2]. Another important category of serum proteins are the apolipopoteins (chylomicrons, Very Low Density Lipoproteins, Low Density Lipoproteins, and High Density Lipoproteins) which are protein-lipid clusters involved in the transport of cholesterol around the body. These lipoprotein complexes have sizes in the range of 8–100 nm, and are well known to be involved in cellular trafficking.

In this in vitro study we begin to shed some light on the detailed nature of the corona of proteins that polyvinyl alcohol microballoons will adopt once they enter biological fluids in the body, bearing in mind that the plasma used in these studies is a model biological solution. We report identification of the full corona of PVA microballoons prepared under different temperature and pH conditions (resulting in different shell thicknesses and microballoon diameters) in human blood plasma and focus on the binding properties of selected proteins from among those identified as being in the microballoon corona. Subtle effects related to protein-protein interactions are highlighted, which may be of significance for the long-term biocompatibility of targeted and theranostic ultrasound contrast agents.

5.2 Experimental details

5.2.1 Human blood plasma

Blood samples were taken from various donors into sample tubes containing EDTA to prevent blood clotting. The samples were centrifuged at 800 g for 5 min. The supernatant was recovered, aliquoted into 1 ml portions, frozen and stored at $-80°C$. Before each experiment, the plasma was thawed and centrifuged again at 16.1×10^3 g for 3 min. The supernatant was used for experiments.

Buffers

10 mM Phosphate, 0.15M NaCl, 1 mM EDTA (PBS buffer) was used as the buffer solution for all protein studies, as it closely resembles physiological conditions (pH, ionic strength, etc.).

Microballoons

Microballoons are produced by rapid stirring of a telechelic Poly(vinyl alcohol) PVA solution in a gas atmosphere by our SIGHT collaborators in Rome [15]. Floating microballoons are separated and dialyzed against water. By varying the parameters (pH and temperature) during polymerization, three different microballoon types are obtained as listed in Table 5.1.

Tissue culture medium

Minimal Essential Medium (MEM) tissue culture medium (Invitrogen Corp.) was supplemented with 10 % fetal bovine serum (FBS, Invitrogen Corp.), 1 % penicillin/streptomycin (Invitrogen Corp.), 1 % L-glutamine (Invitrogen Corp.), and 1 % non-essential aminoacid (Hycrone), and stored at 37°C.

Proteins

The proteins Immunoglobulin G (IgG), Fibrinogen and Human Serum Albumin (HSA) were purchased from Sigma as lyophilized samples and stored at the recommended temperatures.

Table 5.1. Properties of Microballoons

Name	Diameter (nm)	Shell thickness (nm)	Surface type*
MBpH5RT	4.6 ± 0.9	0.6 ± 0.3	A
MBpH5C	2.7 ± 0.5	0.5 ± 0.3	A
MBpH2C	3.0 ± 0.8	0.7 ± 0.3	B

* Note that the surface type is to be understood qualitatively

1D Polyacrylamide gel electrophoesis (PAGE)

SDS/PAGE (Sodium Dodecyl Sulfate Polyacrylamide Gel Electrophoresis) is used to separate proteins by their molecular weight. The detergent SDS (sodium dodecyl sulfate) is added to the protein-microballoon mixture to be investigated, and the protein solution is heated to 99°C for 5 min to denature the proteins. Polyacrylamide gels (12 %) are used to separate the proteins under an electric field such that the negatively charged SDS-protein-micelles are pulled through the gel. Larger proteins encounter more resistance and migrate through the gel slower than smaller proteins. Protein bands are stained using Coomasie blue.

Mass spectrometry

Bands of interest from SDS-PAGE gels (12 %) were excised and digested in-gel with trypsin according to the method of Shevchenko et al. [16]. The resulting peptide mixtures were re-suspended in 0.1 % formic acid and analyzed by electrospray liquid chromatography mass spectrometry (LC MS/MS). An HPLC (Surveyor, ThermoFinnigan, CA) was interfaced with an LTQ ion trap mass spectrometer (ThermoFinnigan, CA). Chromatography buffer solutions (Buffer A, 0.1 % formic acid; Buffer B, 100 % acetonitrile and 0.1 % formic acid) were used to deliver a 72 min gradient (5 min sample loading, 32 min to 40 % Buffer B, 2 min to 80 %, hold 11 min, 1 min to 0 %, hold for 20 min, 1 min flow adjusting). A flow rate of 150 µl/min was used at the electrospray source. Spectra were searched using the SEQUEST algorithm [19] against the indexed uniprot/swiss prot database (http://www.expasy.org; release 3 July 2007). The probability-based evaluation program Bioworks Browser was used for filtering identifications; proteins with Xcorr $(1,2,3) = (1.90, 2.00, 2.50)$ and a peptide probability of $1\ e^{-5}$ or better were accepted.

Isothermal titration calorimetry (ITC)

In order to determine the binding stochiometry of proteins identified from mass spectrometry as binding to the microballoons, solutions of HSA, IgG, and Fibrinogen were titrated into suspensions of microballoons in PBS, and the heat of the reaction recorded over time. Integration of the peaks in the enthalpy diagram yields values of the enthalpy per injected mole of protein. These values can be fitted using models to obtain the binding constants.

The particle stock suspensions (1 mg/ml in water) were diluted to 0.9 mg/ml with 10× PBS, yielding the same concentrations of phosphate, salt and EDTA as in the protein solution titrate. The pH of all sample solutions was adjusted to ensure they coincided within ±0.1. The instrument was rinsed before each experiment using 5 sample cell volumes (SCV) of 2 % Hellmanex cuvette cleaning solution and 10 SCV of MilliQ water. The protein and microballoon solutions were degassed by stirring for five minutes in vacuum, in order to prevent artifacts caused by collapsing gas bubbles in the sample cell.

In a typical experiment, protein (e.g. HSA) was dissolved in PBS to a 40 µM solution as the injectant. The instrument was set to 30 injections, with a volume of 1 µl for the first injection and 5 µl for all the following injections, and an equilibration time of 300 sec after each injection. The stirring speed was set to maximum (300 rpm) to keep the particles in homogenous suspension. The temperature was set to 25°C and the reference power to 5 µcal/sec. The changes in reference power over time were recorded and analyzed using Origin. The data peaks were baseline-corrected, integrated and plotted against the molar ratio of protein and microballoons.

5.3 Results and discussion

5.3.1 Dispersion of the microballoons in PBS buffer

The as-supplied microballoons were at a concentration approximately 1 mg/mL in MilliQ water, and due to their gas-filled nature of the microballoons have high buoyancy, such that they tended to float to the top of the solution with time (within an hour). As proteins are sensitive to pH and ionic strength, it was necessary to perform the binding studies under physiological-type conditions. Thus, we dialyzed the microballons against PBS buffer for several days in order to ensure that the solutions were at the correct pH and salt concentration for the corona determination and protein binding studies, while maintaining the starting concentration of the microballoons.

Sonication of the microballoons for 3 minutes in a standard sonication bath resulted in them taking up buffer into their core and becoming buffer-filled rather than gas-filled. This made working with them easier, since because dispersions were more stable it was possible to separate them using centrifugation. The size of the pellet at the bottom of each flask following centrifugation corresponded well with the shell thicknesses of the respective microballoons, implying that the stability against uptake of buffer into the core (i.e. replacing the gas with buffer) is correlated with the mechanical stability of the microballoons. Light microscopy images (Fig. 5.2) show that sonication did not alter the shape of the microballons, and thus all experiments were carried out on the buffer-filled variants of the microballoons (sonicated microballoons).

5.3.2 Determination of the microballoon protein corona from plasma

Microballoons (1 mg/ml) synthesized at pH 5 at both room temperature and at 5°C were incubated with 450 µl of human blood plasma. After one hour, the sample was sonicated in a sonication bath for 3 min, the particles were pelleted by centrifugation and the supernatant was discarded. The pellet was

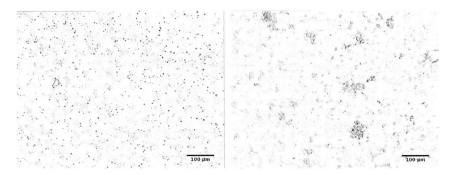

Fig. 5.2. Dark field microscopy images of microballoons after sonication for 3 minutes (left) and following incubation with human serum albumin (right)

washed three times with PBS buffer, saving the supernatant of each step for further investigation. The same protocol was performed for a control experiment except that here the samples were sonicated for 3 minutes immediately after the 1 hour incubation. The pellet and the supernatants of the washing steps were mixed 1:1 with SDS loading buffer and separated using SDS/PAGE.

From Figure 5.3 it is clear that the 3 washing steps are sufficient to remove all of the unbound, or loosely associated proteins from the microballoon corona, leaving only those proteins that are associated with the microballoons for biologically relevant timescales (i.e. timescales whereby molecular recognition could occur, and cellular receptors recognize the presence of the proteins at the microballoon surface). The particles prepared at pH 5 room temperature (MBpH5RT) show a difference in protein band intensity depending on the treatment. The particles that were sonicated before treatment seem to bind more proteins than those who were intact during incubation, likely as a result of the additional surface area provided by the sonicated microballoons, where the internal surface may also be available for binding. For the particles prepared at pH 5 at a cool temperature, (MBpH5C), there is no difference in the protein binding pattern between the two treatments – sonication of the microballlons before of after incubation in plasma did not affect the protein binding pattern.

Microballoons prepared at pH 2 were also studied, and a representative gel is shown in Figure 5.4. In this case, bands were excised from the gel and the proteins were identified using mass spectrometry, and are listed in Table 5.2 according to the band labels shown in Figure 5.4.

The first interesting impression when looking at the peptide list is the high selectivity of the microballoons. From the several thousand proteins in plasma, only a very small fraction can be detected on the microballoons. Of course, this statement needs to be considered with care since the ratio between the most and least abundant proteins in plasma ranges over 10 orders of magnitude whereas the dynamic range of a typical mass spectrometer is only about two orders of magnitude.

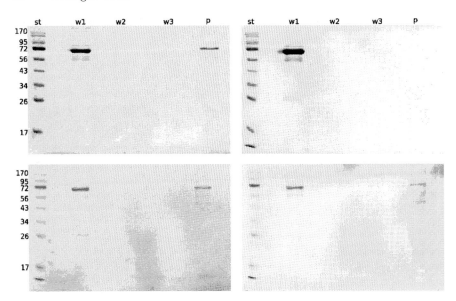

Fig. 5.3. Effects of sonication on microballoon particles prepared at pH 5 Room temperature (top) and at pH 5 cold (bottom). In the gels on the left, the microballoons were sonicated before incubation with plasma, whereas in the gels on the right the microballons were sonicated after incubation with plasma. The supernatants of the washing steps (w1–w3) and the proteins bound to the particles (p) are also shown. The scaling for the mass standard (st) is given in kDa

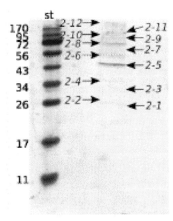

Fig. 5.4. 1D-PAGE of microballoons prepared at pH 2 incubated with plasma for 1 hour, and then separated from unbound proteins by centrifugation and washing 3 times. Bands were excised as shown by the arrows, and the proteins in each band identified by mass spectrometry

Table 5.2. Proteins in the corona of MBs prepared at pH2 as identified by mass spectrometry following excision of the bands illustrated in Figure 5.4

Band	UniProt ID	Name
2–1	P01834	Ig κ chain C region
	P02768	Serum albumin precursor
	P99999	Cytochrome c
2–2	P99999	Cytochrome c
2–3	–	No peptides
2–4	O75636	Ficolin-3 precursor
2–5	P21333	Filamin-A
	P02768	Serum albumin precursor
	P60709	Actin, cytoplasmic 1
2–6	P01834	Ig κ chain C region
	P01857	Ig γ-1 chain C region
	P01859	Ig γ-2 chain C region
	P02768	Serum albumin precursor
	P99999	Cytochrome c
	Q9H4B7	Tubulin beta-1 chain
	P18206	Vinculin
	P02679	Fibrinogen γ chain precursor
2–7	P02675	Fibrinogen β chain precursor
	P02768	Serum albumin precursor
2–8	P02768	Serum albumin precursor
2–9	P01871	Ig μ chain C region
	P02768	Serum albumin precursor
2–10	P18206	Vinculin
	P05106	Integrin β-3 precursor
	P12814	α−actinin-1
	P02768	Serum albumin precursor
	P21333	Filamin-A
2–11	P21333	Filamin-A
2–12	P21333	Filamin-A
	P35579	Myosin-9

5.3.3 Microballoon interaction with individual plasma proteins

Based on the list of proteins identified in Table 5.2 as being part of the protein corona of the microballoons prepared at pH 2, human serum albumin (HSA), fibrinogen and IgG were selected for individual binding studies, using Isothermal Titration Calorimetry to determine the heat of binding and the binding stoichiometry, as carried out previously for polymeric nanoparticles with HSA [10].

5.3.4 Isothermal titration calorimetry

ITC is a very sensitive technique for measuring changes in enthalpy. However, in the case of binding that does not involve enthalpy changes, for example, when only entropic effects are involved, no signal is detected via ITC.

As the gels suggested that the microballoons bind HSA, fibrinogen and IgG, was each was titrated into a solution of microballooons and the heat of reaction measured. All proteins were dissolved at the same weight-per-volume concentration (2.68 mg/ml) and titrated into a microballoon suspension of 0.9 mg/ml in PBS buffer.

As the first injection sometimes carries small air bubbles, the first data point was discarded in every analysis. The data points were fitted using a simple binding model with one binding site and the Levenberg-Marquardt fitting algorithm. Since the full binding curve was not recorded, all parameters carry significant errors. The results of the protein binding studies are shown in Figure 5.5.

From Figure 5.5 it is clear that the different proteins display quite different behavior upon interaction with the microballoons. HSA shows a typical saturation-binding curve whereas the other protein behavior is more complex.

Despite the difficulties with performing the measurements, and the fact that the full binding curves could not be measured, an attempt was made to fit the data, and obtain the binding constants for each of the proteins to the microballons. From the HSA binding curve in Figure 5.5 (left), it is clear that the binding of HSA to the microballoons is an exothermic process, which

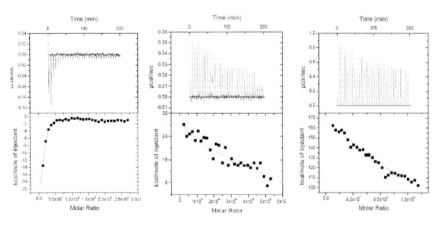

Fig. 5.5. Isothermal titration calorimetry of 0.9 mg/ml microballoons (prepared at pH 5 room temperature) in PBS with 5 µl injections of 2.68 mg/ml solutions of the different proteins. (Left) HSA; (Center) Fibrinogen; (Right) IgG

results in output of heat. The binding constant was found to be K = (2.3 ± 1.1) × 10^6 M^{-1}, although the fit is not very reliable since the binding stochiometry carried errors.

In the case of fibrinogen, the molar enthalpy change was approximately in the same order of magnitude as for HSA. The sign of the signal, however is different, indicating that the binding of fibrinogen to the microballoons is endothermic. Because of fibrinogen's significantly higher mass (340 kDa) compared to HSA (69 kDa), the molar ratios are quite different for the binding of the two proteins to the microballoons. The binding constant for fibrinogen was determined to be K = (9.9 ± 8.0) × 10^{-5} M^{-1}.

Titration of IgG into the particles yields huge binding enthalpy changes as seen in Figure 5.5 (right). As the binding curve could not be fully recorded, a fit was not possible. The significantly higher enthalpy changes compared to fibrinogen, around 160 kcal/mol vs. around 24 kcal/mol in the first few injections, rendered the binding highly endothermic.

5.4 Discussion

We conceive of the proteins associated with the microballoons possessing a very wide range of affinities for the particle surface. In essence we expect a huge range of equilibrium constants (one for each protein) representing the quite different (and competitive) binding mechanisms present. This means that we see the proteins associated to the microballoons as a 'corona', rather than a solid fixed layer. This corona may not immediately reach equilibrium when exposed to a biological fluid. Proteins with high concentrations and high association rate constants (such as albumin) will initially occupy the microballoon surface but may also dissociate quickly to be replaced by proteins of lower concentration, slower exchange and higher affinity. The biological outcome may also differ depending on the relative protein exchange between microballoon surface and cellular receptors.

Following the proteomic assessment of the protein corona of the microballoons prepared at pH 2, the presence of HSA in all bands is not surprising because of its high abundance in plasma and the large size of the microballoons which resembles a flat surface (low surface curvature). For proteins at such high concentrations, the resolving power of SDS/PAGE is not enough. Most of the proteins identified on the microballoons belong to the 22 most abundant proteins that make up 99 % of plasma [8]. Identified proteins that do not belong to this list are mostly related to the cytoplasm or cytoskeleton (Ficolin-3, Filamin-A). These proteins are, however, also found in the blood plasma in low concentrations under natural circumstances. In our experiment, the concentration of these proteins might be slightly increased due to cells that ruptured during plasma purification. This has potentially quite significant consequences for our understanding of polymeric biomaterials, many of which are traditionally considered to be essentially non-binding for proteins [11].

Nevertheless, the detection of these proteins on the microballoons indicates a strong specificity of the PVA microballoons for lower abundance proteins, and a lower affinity for the higher abundance proteins such as albumin. This is despite the fact that albumin is shown to bind, both by the fact that it is identified proteomically in the microballoon corona, but also by the ITC binding experiments. However, its affinity is low, and it is replaced by higher affinity lower abundance proteins over time, as has also been shown for a range of different nanoparticles [4].

The presence of many actin-binding proteins such as filamin-A, vinculin, α-actinin, and integrin is very interesting, as these proteins are involved in cell adhesion. This suggests that the microballoon surface will present a favorable binding surface for cells, as would be expected based on its composition (PVA) and as is known for PVA hydrogels [20]. This also has deep implications for targeting, and delivery, as it suggests that the microballoons may be able to anchor to specific cells, and remain in situ for a sufficient time to deliver therapies and to image disease regression in response to the therapy. Additionally, the presence of the protein talin in the corona suggests a strong role for protein-protein binding, as talin is capable of binding to the actin cytoskeleton by binding vinculin or integrin [6]. Thus, it may be that binding of some of the identified proteins is not a consequence of interaction with the microballoons at all but is rather a consequence of protein-protein interactions. In this case, certain proteins that genuinely only have low binding affinities to the microballoons could be detected since they bind to other proteins that are bound to the microballoons. From the general viewpoint of identifying the proteins in the microballoon protein corona, this does not matter in terms of characterization of the overall composition of the microballoon protein corona. However, from the viewpoint of obtaining a deeper assessment of the potential long-term biocompatibility of microballoons as contrast agents, the role of all identified proteins and their biological impact is necessary.

5.5 Conclusions

This study provided a general characterization of the interaction of poly(vinyl alcohol) microballoons, which are being researched within the SIGHT consortium for use as theranostic contrast agents, with some of the major plasma proteins. Microballoons prepared under different temperature and pH conditions resulting in different shell thicknesses, diameters and aldehydic surface characteristics, were investigated, and the composition of their protein coronas identified. Binding properties of selected proteins from the corona were investigated using isothermal titration calorimetry.

Very interesting binding patterns emerged from the full proteomic studies, and quite high selectivity and specificity of protein binding was observed. A role for protein-protein interaction in determining the composition of the corona is also highlighted. To investigate this further, ITC could also be applied under competitive binding conditions, for example where the particles are already coated with one type of protein and another is titrated in, to see if protein-protein interactions play a significant role. Such studies are ongoing within the SIGHT project.

From this preliminary study, it is clear that a complete understanding of the protein corona and its evolution and dynamics in plasma is necessary in order to design contrast agents which do not affect their host, do not induce an inflammatory response, and can circulate in the blood stream with sufficient longevity, and target the desired tissue with sufficient specificity, to function effectively as theranostic agents.

Acknowledgements

This work was funded by the FP6 IST project SIGHT (IST-2005-033700). Microballoons were provided by Gaio Paradossi and Mariarosaria Tortora from University Rome Tor Vergata. ITC experiments were performed at the Department of Biophysical Chemistry, Lund University, Sweden.

References

1. Anderson JM (2001) Biological responses to materials. Annu Rev Mater Res 31:81–110
2. Anderson NL and Anderson NG (1977) High Resolution Two-Dimensional Electrophoresis of Human Plasma Proteins. Proceeding of the National Academy of Sciences 74:5421–5425
3. Cavalieri F, El Hamassi A, Chiessi E and Paradossi G (2007) Tethering Functional Ligands onto Shell of Ultrasound Active Polymeric Microbubbles. Submitted to Bioconjugate Chem
4. Cedervall T, Lynch I, Foy M, Berggård T, Donelly S, Cagney G, Linse S and Dawson KA (2007) Detailed Identification of Plasma Proteins Absorbed to Copolymer Nanoparticles. Angewandte Chemie Int Ed 46:5754–5756
5. Cedervall T, Lynch I, Lindman S, Nilsson H, Thulin E, Linse S and Dawson KA (2007) Understanding the nanoparticle protein corona using methods to quantify exchange rates and affinities of proteins for nanoparticles. PNAS 104:2050–2055
6. Critchley DR (2000) Focal adhesions – the cytoskeletal connection. Current Opinion in Cell Biology 12:133–139
7. Doillon CJ and Cameron K (1990) New approaches for biocompatibility testing using cell culture. Int J Artif Organs 13:517–520
8. Issaq HJ, Xiao Z and Veenstra TD (2007) Serum and plasma proteomics. Chem Rev 107:3601–3620
9. Klein J (2007) Probing the interactions of proteins and nanoparticles. PNAS 104:2029
10. Lindman S, Lynch I, Thulin E, Nilsson H, Dawson KA and Linse S (2007) Systematic investigation of the thermodynamics of HSA adsorption to N-isopropylacrylamide-N-tert-butylacrylamide polymeric nanoparticles. NanoLetters 7:914–920
11. Liu SX, Kim J-T and Kim S (2008) Effect of Polymer Surface Modification on Polymer-Protein Interaction via Hydrophilic Polymer Grafting. Journal of Food Science 73:E143–E150

12. Lynch I, Dawson KA and Linse S (2006) Detecting crytpic epitopes in proteins adsorbed onto nanoparticles. Science STKE 327:14
13. Muthusamy B, Hanumanthu G, Suresh S, Rekha B, Srinivas D, Karthick L, Vrushabendra BM and Sharma S et al. (2005) Plasma Proteome Database as a resource for proteomics research. Proteomics 5:3531–3536
14. Ong ATL, Hoye A, Aoki J, van Mieghem CAG, Rodriguez Granillo GA, Sonnenschein K, Regar E, McFadden EP, Sianos G, van der Giessen WJ, de Jaegere PPT, de Feyter, P. van Domburg RT and Serruys PW (2005) Thirty-day incidence and six-month clinical outcome of thrombotic stent occlusion after bare-metal, sirolimus, or paclitaxel stent implantation. J Am Coll Cardiol 45:947–953
15. Paradossi G, Cavalieri F, Chiessi E, Spagnoli C and Cowman MK (2003) Poly(vinyl alcohol) as versatile biomaterial for potential biomedical applications. J Mat Sci Materials in Medicine 14:687–691
16. Shevchenko A et al. (1996) Mass Spectrometric Sequencing of Proteins from Silver-Stained Polyacrylamide Gels. Anal Chem 68(5):850–858
17. Williams D (2007) Metastable Biocompatibility: A New Approach. Medical Device Technology
18. Wilson CJ, Clegg RE, Leavesley DI and Pearcy MJ (2005) Mediation of biomaterial-cell interactions by adsorbed proteins: a review. Tissue Eng 11:1–18
19. Yates JR et al. (1995) Method to Correlate Tandem Mass Spectra of Modified Peptides to Amino Acid Sequences in the Protein Database. Anal Chem 67(8):1426–1436
20. Zajaczkowski MB, Cukierman E, Galbraith CG and Yamada KM (2003) Cell–Matrix Adhesions on Poly(vinyl alcohol) Hydrogels. Tissue Engineering 9:525–533

Chapter 6

Polymer Based Biointerfaces: A Case Study on Devices for Theranostics and Tissue Engineering

Pamela Mozetic, Mariarosaria Tortora, Barbara Cerroni, and Gaio Paradossi

6.1 Introduction

In the design of new devices supporting biomedical applications, the focus on the processes occurring at the interface with cells is a major issue. In this context, our interest on the formulation of new biomaterials and on the synthesis of next-generation ultrasound contrast agents (UCAs) lead us to the in vitro study to assess the biocompatibility of these novel devices. As UCAs are designed for parenteral administration, the time response of the primary immune system, mainly macrophages, should be addressed.

Our UCA [2], a poly (vinyl alcohol), PVA, based microbubble, has been designed as an injectable multifunctional device suitable for diagnostics as well as therapeutic treatment promoted by insonification. PVA is a polymer already used for biomedical applications. However, in contrast to other PVA based biomaterials, we have crosslinked modified PVA chains [1] carrying out a one-pot reaction in aqueous medium without introducing external crosslinkers that could in principle jeopardize the biocompatibility of the starting PVA material (see also chapter "Design of novel polymer shelled ultrasound contrast agents: toward a ultrasound triggered drug delivery."). This reaction can be carried out without stirring the reaction medium or foaming the medium by applying a high shear rate stirring. In the first case a thin hydrogel membrane is obtained whereas an aqueous dispersion of microballoons, MBs, is the final product of the latter case. The polymer shell of the microballoons is expected to have the same chemical features as the membrane hydrogels. For this reason we have addressed our attention to biocompatibility studies of both hydrogels and microballoon systems as biointerface assessment is key in any application involving biomedical applications. We addressed in vitro biocompatibility study toward fibroblasts and macrophages as cell types having direct interaction with polymer surfaces in the blood stream.

When studying the processes occurring between microballoons and cells, the establishment of the biointerface is a major issue as the floating tendency of microballoons does not allow effective contact with cells. In order to

Paradossi, G., Pellegretti, P., Trucco, A. (Eds.)
Ultrasound contrast agents. Targeting and processing methods for theranostics
© Springer-Verlag Italia, 2010

accomplish such contact, all polymeric microballoons were transformed into microcapsules by equilibrating them in ethanol followed by reconditioning in PBS buffer. In this way the air-filled microballoons were converted into sterile microcapsules having the same dimensions as the starting MBs [3]. Due to the increased density of the new particles with a core containing PBS, the microcapsules can be brought into contact with cells easily. This choice was dictated by the need to accomplish close contact of the cells with the sterile polymer devices, assuming that the chemical and physical surface features of the microcapsules are maintained during the microballoon-to-microcapsule conversion.

6.2 Materials and methods

6.2.1 Reagents

Dulbecco's modified eagle medium (DMEM), L-glutamine 200 mM and penicillin/streptomycin solution (10000 U/ml and 10 mg/ml, respectively) were obtained from HyClone. Fetal bovine serum, Giemsa staining, MTT (3-(4,5-dimethylthyazol-2-yl)-2,5-diphenyltetrazolium bromide) and phosphate buffered saline (PBS) and HBSS were purchased from Sigma Aldrich. Live/Dead viability/cytotoxicity kit[®] was obtained from Molecular Probes, Invitrogen.

6.2.2 PVA hydrogels and MBs biocompatibility on NIH3T3 fibroblast mouse cell line

MBs fabricated at pH5 and room temperature (MBpH5RT), at pH5 and at 4°C (MBpH5C) and at pH2 and at 4°C (MBpH2C) were used throughout this study. MBs were functionalized with Arg-Gly-Asp, RGD, tripeptide as described in the chapter "Design of novel polymer shelled ultrasound contrast agents: toward a ultrasound triggered drug delivery". All types of MBs were sterilized by EtOH 70 % and re-suspended in sterile PBS buffer.

6.2.3 NIH3T3 fibroblasts and RAW 264.7 macrophage cultures

NIH3T3 mouse fibroblasts from Istituto Sperimentale Zoprofiattico della Lombardia e dell'Emilia Romagna (Italy), were maintained in Dulbecco's modified Eagle medium (DMEM) contaning 10 % FBS, 1 % penicillin/streptomycin solution and 1 % glutamine at 37°C in a 5 % CO_2 atmosphere. Cell viability and cell proliferation were determined by MTT assay [4].

RAW 264.7 macrophages were purchased from the Istituto Zooprofilattico Sperimentale della Lombardia ed Emilia Romagna (Italy). The cells were maintained in complete medium consisting of DMEM supplemented with 2 mM L-glutamine, 10 % heat-inactivated fetal bovine serum and 1 % antibiotic mixture, consisting of 100 U/ml of penicillin and 100 µg/ml of streptomycin.

Macrophages were cultured in a humidified 5 % CO_2 atmosphere at 37°C. The macrophages were plated at the desired concentration in 24 well plates and left to adhere for 12 hours before use.

6.2.4 Microballoon suspension sterilization

MB suspension at a concentration of 1 mg/ml, determined by dry weight analysis, was sterilized in EtOH 70 %, and then transferred into the specific cell medium.

6.2.5 NIH3T3 fibroblast and RAW 264.7 macrophage viability in the presence of PVA hydrogels, RGD modified PVA-hydrogels and PVA based MBs

Fibroblasts were plated in serum free DMEM at a density of 50,000/well on hydrogel surface or incubated with different amounts of MBs in serum free DMEM (25 and 50 µg/well). Macrophages were plated in complete medium at a density of 200,000/well and left to adhere for 12 hours. Subsequently cells were washed and incubated with different amounts of MBs (25 and 50 µg/well) for 4, 8 and 18 hours using serum free DMEM in the presence of MTT reagents. At the end of the incubation period the medium was discarded and formazan crystals dissolved in DMSO. Formazan concentration was determined by measuring the absorbance of the fomazan in DMSO solution at 570 nm.

6.2.6 NIH3T3 fibroblast and RAW264.7 macrophage proliferation on PVA hydrogels and after exposure to PVA based MBs

PVA hydrogels and RGD-PVA hydrogels were rinsed three times, first with PBS and then with complete medium. NIH3T3 fibroblasts and RAW 264.7 macrophages were seeded at a density of 50,000/well on PVA hydrogels and RGD-PVA hydrogels or incubated with different amounts of MBs in complete medium (25 and 50 µg/well). Macrophages were plated in complete medium at a density of 200,000/well and left to adhere for 12 hours. Cell proliferation was determined by MTT assay [4] after 1, 3 and 7 days, as previously described in section 6.2.5.

6.2.7 Phagocytosis of Microballoons

Adhering RAW 264.7 cells (500,000 cells/well) in 12-well plates were incubated for 4, 8 and 18 hours with 50 µg/well of MBs in complete medium. After incubation, the medium was discarded and cells with non-internalized MBs were washed off with PBS. Cells were fixed with methanol for 10 minutes and stained with Giemsa for 20 minutes. Excess dye was washed off with water, and the wells were dried and observed by light microscopy.

6.2.8 RAW 264.7 viability by live/dead assay

A live/dead staining kit® (Molecular Probes) is a two-color fluorescence assay for animal cell viability. This kit provided two probes, SYTO 10® (green fluorescent nucleic acid stain) and DEAD Red (ethidium homodimer-2 nucleic acid stain). SYTO 10® is a highly membrane-permeant dye and it labels all cells, including those with intact plasma membrane. DEAD Red is a cell-impermeant red fluorescent nucleic acid stain that labels only cells with compromised membrane.

Approximately 200,000 macrophages were seeded onto coverslips and left to to adhere in complete medium for 12 hours. Macrophages were treated with 50 μg/slide of different types of MBs. Following 1, 3 and 7 days of treatment, adherent cells on the coverslips were washed with HBSS and incubated with dye mixture (as suggested by manufacturer). After staining, cells were washed with HBSS and fixed with glutaraldehyde 4 % at least 15 minutes before observation.

Viable cells, observed under a confocal laser scanning microscopy (CLSM) at a magnification of 60 x, appeared green whereas necrotic cells presented a red color. Seven microscopic fields in each coverslip were chosen to determine the percentage of live/dead cells in experimental and control groups.

6.2.9 MB internalization studied by confocal laser scanning microscopy (CLSM)

For CLSM observations, RAW 264.7 were stained with fluorescein-labeled phalloidine (fluorescein isothiocyanate labeled phalloidin, Sigma), to visualize the actin cytoskeleton of the macrophages treated with rhodaminated microballoons. Cells were treated with different types of rhodaminated microballoons for 24 hours to induce MB uptake. Then cells were washed with PBS and fixed for 5 minutes in 3.7 % formaldehyde in PBS at room temperature. After fixation, the macrophages were washed extensively in PBS and incubated for 40 minutes with 0.1 % Triton X-100 to permeabilize the cell membrane. FITC-phalloidine, at a concentration of 15 μM in PBS, was added and samples were incubated over night at 4°C. The samples were washed several times with PBS and mounted on coverslips for microscope evaluation.

6.3 Results and discussion

PVA based biomaterials as hydrogels and microballoons need an assessment of their viability features. Both devices are subjected to structural variations depending on the different synthetic conditions. In particular polymer shelled MBs can be obtained with different average size and shell thickness depending on pH and temperature preparation conditions. These structural differences may correspond to a change in the surface morphology, a factor influencing the interactions pattern with living material. As a case study on the cellular responses to both materials, two cell lines, NIH 3T3 mouse fibroblast and RAW 264.7 mouse macrophages, were chosen in consideration of the potential use of PVA based hydrogels and microballoons as scaffolds for tissue engineering and as injectable devices for theranostics, respectively. PVA has good chemical versatility due to the presence of a hydroxyl moiety on the backbone. This feature allows the grafting of several biologically relevant molecules. In this context, we studied the in vitro biocompatibility of hydrogels and microballoons based on PVA and containing the tri-peptide, Arg-Gly-Asp, RGD. The RGD sequence is the ligand to the attachment site of a large number of proteins involved in the adhesive processes occurring in the extracellular matrix, blood, and cell aggregation [5]. Considering the use of PVA as starting material for a potential scaffold for controlled cellular growth, and the design of MBs as targeted in situ delivery systems, the conjugation with RGD was carried out on both PVA hydrogels and microballoons and their biocompatibilities were compared with the un-derivatized devices.

For the preparation of the unmodified PVA hydrogels and MBs, as well as the corresponding RGD modified devices, we refer to the chapter "Design of novel polymer shelled ultrasound contrast agents: toward a ultrasound triggered drug delivery" of this book.

The biocompatibility of the RGD-modified and unmodified devices was assessed using fibroblasts according to the MTT assay [4] (see Fig. 6.1).

The proliferation rate of NIH3T3 fibroblasts was monitored by MTT assay after 1, 3 and 7 days, showing that the cells' metabolism, seeded on PVA based hydrogels or in contact with PVA shelled MBs, was not influenced by

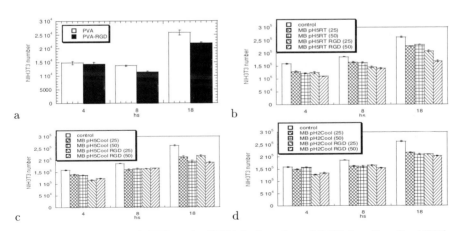

Fig. 6.1. Cytotoxicity (MTT test) of PVA hydrogels and RGD functionalized PVA hydrogels (**a**); cytotoxicity (MTT test) of non-modified and RGD modified PVA based MBs: (**b**) MBpH5RT, (**c**) MBpH5C and (D) MBpH2C on NIH3T3 fibroblasts. In parenthesis the amounts (µg/well) of MBs are indicated

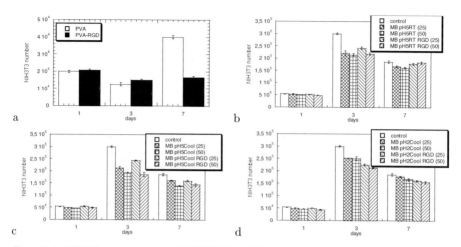

Fig. 6.2. MTT measurement of NIH3T3 proliferation rate after exposure to PVA hydrogels (**a**) or PVA based microballoons (**b**, **c**, **d**). In parenthesis the amounts (μg/well) of MBs are indicated

the presence of these materials (see Fig. 6.2). On the basis of these findings, we could assess the biocompatibility of non-modified or RGD modified devices. The amount of RGD bound on the surface of the MBs was determined by BCA colorimetric assay [6] resulting in 5×10^{-8} RGD moles per mg of MB (details in "Design of novel polymer shelled ultrasound contrast agents: toward a ultrasound triggered drug delivery" of this book).

The primary immunoresponse will be triggered once MBs are injected into the blood stream with the attack of neutrophills against the exogenous particles. Therefore in vitro study of macrophage response to the presence of unmodified and RGD functionalized MBs was addressed in view of parenteral administration of PVA shelled MBs. In particular we studied the internalization of PVA based MBs and of the RGD modified PVA MBs with murine RAW 264.7 macrophages. After 4 hours phagocytosis was not yet started, whereas after an incubation of 8 hours, the internalization process had clearly taken place, as shown in Figure 6.3.

Phagocytosis was followed by confocal laser scanning microscopy (CLSM) assessment of the intracellular localization of the microparticles in the cytoplasm. The morphology of the RAW 246.7 macrophages during the process was visualized by staining the cytoskeleton actin with fluorescein-labeled phalloidine (green), while the microparticles were labeled with rhodamine (red) (see Fig. 6.4).

Questions as to whether and how MBs influence macrophage viability during phagocytosis were addressed by carrying out the live/dead assay on cells incubated with the polymeric shelled particles and comparing the results with a control experiment where the cells were cultured in medium in the absence of particles. Inspection by fluorescent microscopy showed that cell

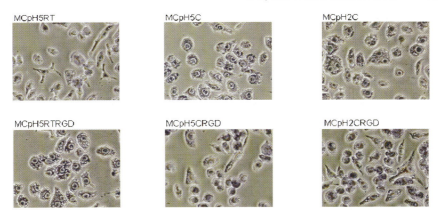

Fig. 6.3. Giemsa staining of RAW 264.7 after 8 hours of incubation with different types of MBs

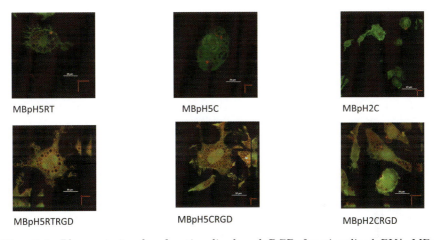

Fig. 6.4. Phagocytosis of unfunctionalized and RGD functionalized PVA MBs by RAW 264.7 macrophages CLSM. Actin cytoskeleton was labeled with FITC-phalloidin (green) and microparticles were labeled with TRITC (red). Optical sections (xy, left panel) and yz projections (right panel) allow extra-cellular and internalized microparticles to be distinguished

viability was independent from the type of MBs used and did not change with time (see Fig. 6.5). Table 6.1 summarizes the results of this study after 1, 3 and 7 days of macrophage exposure to MBs, indicating that RAW 246.7 viability is not influenced by the presence of MBs and by the exposure time to MBs.

Evaluation of cytotoxicity of MBs and RGD modified MBs was carried out by MTT test during exposure of RAW 246.7 macrophages to the micropartices (see Fig. 6.6) as MB internalization during this time can have an impact on the metabolic pathways of phages. The results shown in Figure 6.6 indicate

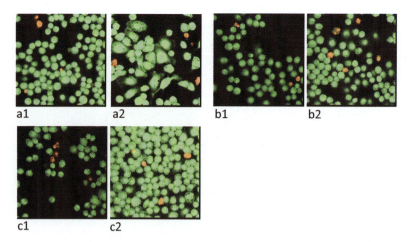

a1   a2   b1   b2

c1   c2

Fig. 6.5. Live/Dead assay of treated RAW 264.7 with different types of MBs at day 7; live cells were green and dead cells were red. a1) cells treated with MCpH5RT; a2) cells treated with MCpH5RTRGD; b1) cells treated with MCpH5C; b2) cells treated with MCpH5CRGD; c1) cells treated with MCpH2C; c2) cells treated with MCpH2CRGD

Table 6.1. RAW 264.7 viability after 1, 3 and 7 days exposure to different types of PVA based microballons. Cell viability was expressed as a percentage of live cells (green figures) vs. the percentage of dead cells (red figures)

Treatment	after 1 day	after 3 days	after 7 days
Control	100/0	91/9	92/8
MBpH5RT	91.5/8.5	92/8	89.5/10.5
MBpH5C	85/15	89/11	89/11
MBpH2C	85/15	93/7	91/9
MBpH5RTRGD	87/13	90/10	87/13
MBpH5CRGD	91/9	89/11	85/15
MBpH2CRGD	97/3	91/9	88/12

a limited decrease of the metabolic activity of macrophages monitored at 4, 8 and 18 hours after incubation with MBs and RGD modified MBs.

Cell proliferation rate is another key parameter in macrophage viability study. MTT assay on RAW 264.7 cells showed a proliferation rate decrease during the first 3 days of exposure to MB5RT. This effect is enhanced in the presence of RGD-modified MB5RT, probably due to a more favourable adhesion of MBs to the cell surface. On the other hand, the interaction of macrophages with MB2C and MB5C seems to leave unaffected the proliferation rate of RAW 247.7 macrophages (see Fig. 6.7).

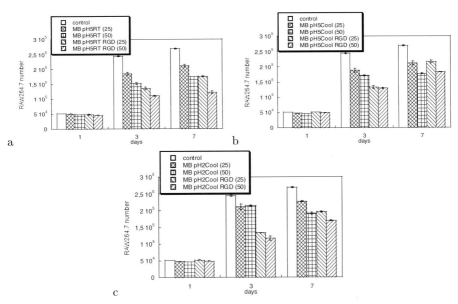

Fig. 6.6. MTT test: RAW 264.7 macrophage viability after incubation with different amounts and types of microballoons: (**a**) MBpH5RT ± RGD; (**b**) MB pH5C ± RGD; (**c**) MBpH2C ± RGD. Bars represent mean values ± SD (n = 4). In parenthesis the amounts of MBs (µg/well) are indicated

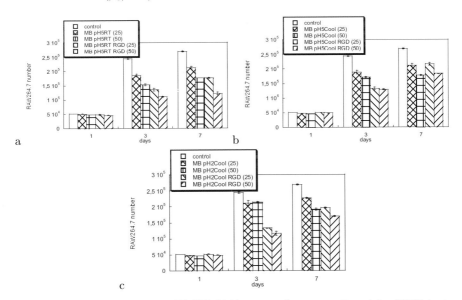

Fig. 6.7. Proliferation rate of RAW 264.7 macrophages monitored by MTT test: (**a**) MBpH5RT ± RGD; (**b**) MBpH5C ± RGD; (**c**) MBpH2C ± RGD. Bars represent mean values ± SD (n = 4). In parenthesis the amounts of MBs (µg/well) are indicated

6.4 Conclusions

In this study we have presented novel results concerning fibroblast and macrophage viability when exposed to PVA based polymer hydrogels designed to function as scaffolds for tissue engineering and microballoons aimed at theranostics applications. Both devices were used in the unmodified form and as RGD-modified materials.

In all cases cell viability was assessed, indicating that the PVA based materials are biocompatible and can be suitably used in the development of next generation ultrasound contrast agents and scaffolds. In the case of phagocytosis, RGD promotes interaction at the biointerface increasing the efficiency of the internalization process. Macrophage activity in the presence of microparticles is evident after about 8 hours. This is a suitable time lag for not having any microballoon depletion in the systemic circulation due to immune system response.

Ongoing studies on the processes occurring at the biointerface are focussing on the ultrasound assisted therapeutic treatment of cancer. To this aim, new functionalization of PVA shelled microballoons will involve the surface modification of the particle with the ligand of the CD44 receptor, over-expressed by tumor cells, and loading with anticancer specific drugs.

MBs can be envisaged as therapeutic gas carriers. Nitric oxide (NO) plays a central role in controlling arterial thrombosis and in cardiovascular diseases by inhibiting the platelet aggregation process. This process is regulated by giving a deactivating signal for protein membrane integrins, the major platelet adhesion receptors. The localized production of NO, naturally occurring in arterial vessels, is carried out by the NO synthase enzymatic system. Inhibition of platelet aggregation in the coagulation cascade process is due to the antagonist action of NO against integrin-fibrinogen induced platelet adhesion. However, sometimes the natural supply of NO is not sufficient to prevent clotting. The design of devices for suitable transport and delivery of NO is therefore important. Due to their structure, microballoons are intrinsically suitable for carrying gasses. We are addressing this feature to perform NO release by means of polymer shelled microballoons [3]. NO release can theoretically be concentrated in vessels with acute thrombosis by bursting of the microparticles upon insonification. We demonstrated that PVA shelled microballoons can be loaded with NO and that the colloidal stability of these NO loaded bubbles can in principle enable local delivery when the particle shell has a suitable flexibility.

References

1. Barretta P, Bordi F, Rinaldi C and Paradossi G (2000) A Dynamic Light Scattering Study of Hydrogels Based on Telechelic PVA. J Phys Chem B 104:11019–11026

2. Cavalieri F, El-Hamassi A, Chiessi E and Paradossi G (2005) Stable polymeric microballoons as multifunctional device for biomedical uses: synthesis and characterization. Langmuir 21:8758–8764
3. Cavalieri F, Finelli I, Tortora M, Mozetic P, Chiessi E, Polizio F, Brismar TB and Paradossi G (2008) Polymer Microballoons As Diagnostic and Therapeutic Gas Delivery Device. Chem Mater 2:3254–3258
4. Mosmann T (1983) Rapid colorimetric assay for cellular growth and survival: application to proliferation and cytotoxicity assays. Journal of immunological methods 65:55–63
5. Ruoslahti E (1996) RGD and other recognition sequences for integrins. Ann Rev Cell Dev Biol 12:697–715
6. Stoscheck CM (1990) Quantitation of Protein. Methods in Enzym 182:50–69

Chapter 7

Ultrasound Contrast Agent Microbubble Dynamics

Marlies Overvelde, Hendrik J. Vos, Nico de Jong, and Michel Versluis

Abstract. Ultrasound contrast agents are traditionally used in ultrasound-assisted organ perfusion imaging. Recently the use of coated microbubbles has been proposed for molecular imaging applications where the bubbles are covered with a layer of targeting ligands to bind specifically to their target cells. In this chapter we describe contrast agent microbubble behavior starting from the details of free bubble dynamics leading to a set of equations describing the dynamics of coated microbubbles. Experimentally, the dynamics of ultrasound contrast agent microbubbles is temporally resolved using the ultra-high speed camera Brandaris 128. The influence of a neighboring wall is investigated by combining the Brandaris camera with optical tweezers. It was observed that the presence of the wall can alter the bubble response. A detailed description of the bubble-wall interaction may therefore lead to improved molecular imaging strategies.

7.1 Introduction

Ultrasound is widely used in medical imaging in gynecology, cardiology, radiology and urology. In cardiology for example ultrasound imaging is used to assess the heart wall motion and the heart valves. Ultrasound Contrast Agents [22] (UCA) are used to enhance endocardial border delineation and to assess perfusion. UCAs consist of small microbubbles with a radius of 1 to 5 μm. The bubbles have a high scattering cross section because of their compressibility giving rise to a strong echo. The microbubbles are coated with a phospholipid monolayer, a polymeric shell or proteins to reduce the capillary pressure and to prevent them from dissolving in the blood, see Figure 7.1.

An emerging application is the use of these type of microbubbles in molecular imaging. Here the bubbles contain targeting ligands which bind specifically to their target cells, for diagnosis at a cellular level. For such molecular imaging applications it is desirable to distinguish freely floating microbubbles from targeted ones.

While the response of a bubble depends on the applied acoustic pressure, it also depends strongly on the applied ultrasound frequency. At low acoustic

Paradossi, G., Pellegretti, P., Trucco, A. (Eds.)
Ultrasound contrast agents. Targeting and processing methods for theranostics
© Springer-Verlag Italia, 2010

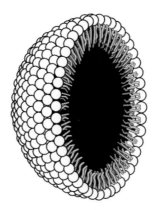

Fig. 7.1. Schematic drawing of a microbubble with a phospholipid monolayer

driving the bubble response is maximum at its resonance frequency. The resonance frequency of microbubbles with a radius of 1–5 μm is in the megahertz
range, which nicely (and for obvious reasons) coincides with the optimum
imaging frequencies used in medical ultrasound imaging. For higher acoustic
pressures the microbubbles show strong non-linear behavior, producing higher
harmonics. Most contrast-enhanced ultrasound imaging techniques, such as
power modulation [3] and pulse inversion [19], exploit the second harmonic
signal to distinguish tissue from contrast.

A targeted microbubble is a complex system containing a gas core, a shell
and targeting ligands. To understand the behavior of these targeted microbubbles we need to understand the dynamics of a gas bubble and the influence of
its shell and targeting ligands. There exists a full theoretical and experimental understanding of the behavior of uncoated gas bubbles in the mm-range
in free space [15, 17, 21, 24, 30, 31, 33, 35] and some of these models are also
used to describe the bubble behavior in the μm-range. The validation of these
models in the μm-range has been limited because of experimental complexity,
predominantly because of the rather short lifetime of the uncoated bubbles.
Several theoretical models have been developed for bubbles with a viscoelastic
coating [6, 7, 18, 36]. Recently, these models were extended to account also for
buckling and rupture of the shell [26]. Acoustical characterization of bubble
suspensions and acoustical and optical experiments for example with the Brandaris 128 camera [5], see Figure 7.2, performed on single bubbles confirm the
physical picture governed by the visco-elastic properties of the shell [16,26,38].
The change in dynamics between untargeted and targeted microbubbles has
been investigated by Zhao et al. [40, 41] and Lankford et al. [23].

In general, experiments on single UCA microbubbles are performed on
bubbles floating against a (capillary) wall. This may not represent the ideal
comparison to theory and numerical simulations, as in the models the bubble
is assumed to reside in an infinite medium. Garbin at al. [14] used an optical
tweezers setup to manipulate single UCA microbubbles and it was shown

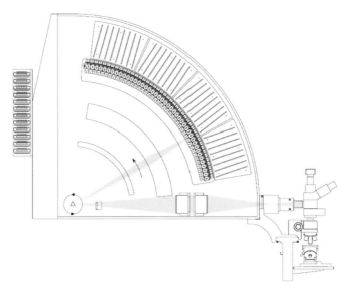

Fig. 7.2. Schematic drawing of the Brandaris 128 camera. The rotating mirror sweeps the light beam projecting the microscope image on the CCD's. The mirror sweeps the image over the CCD's with a minimum interframe time of 40 ns or equivalent a framerate of 25 Mfps

experimentally that the UCA microbubbles' dynamics is hampered by the presence of a wall. A theory on the influence of a rigid wall exists for uncoated gas bubbles and it includes translational oscillations [9, 27]. The influence of the wall on the bubble response may prove to be useful for the discrimination of targeted and freely floating microbubbles in molecular imaging applications.

In the following sections the current understanding of the dynamics of single UCA microbubbles, both theoretically and experimentally, is discussed. Section 7.2 starts with a description of the behavior of an uncoated bubble using the non-linear Rayleigh-Plesset equation. Through linearization we derive an expression for the resonance curve, the resonance frequency and the damping. In section 7.3 we describe in more detail the influence of a viscoelastic shell, while the influence of a rigid wall on the dynamics of microbubbles is discussed in section 7.4. The chapter is concluded with a discussion and conclusion in section 7.5.

7.2 Dynamics of a free gas bubble

The dynamics of an uncoated bubble in free space was first described by Lord Rayleigh [35] and was later refined by Plesset [31], Noltingk & Neppiras [29,30] and Poritsky [32] to account for surface tension and viscosity of the liquid. A popular version of the equation of motion describing the bubble dynamics (often referred to as *the* Rayleigh-Plesset equation) is given by:

$$\rho\left(\ddot{R}R + \frac{3}{2}\dot{R}^2\right) =$$

$$\left(P_0 + \frac{2\sigma_w}{R_0}\right)\left(\frac{R_0}{R}\right)^{3\kappa}\left(1 - \frac{3\kappa\dot{R}}{c}\right) - P_0 - P_a(t) - 4\nu\rho\frac{\dot{R}}{R} - \frac{2\sigma_w}{R}, \quad (7.1)$$

where ρ is the liquid density, ν the kinematic viscosity of the liquid, c the speed of sound in the liquid, σ_w the surface tension of the gas-liquid system and κ the polytropic exponent of the gas inside the bubble. P_0 is the ambient pressure and $P_a(t)$ the applied acoustic pressure. R_0 is the initial bubble radius, R represents the time-dependent radius of the bubble, while $\dot{R}$ and $\ddot{R}$ represent the velocity and the acceleration of the bubble wall, respectively. The bubble is assumed to be surrounded by an infinite medium and it remains spherical during oscillations. The bubble radius is small compared to the acoustic wavelength. The gas content of the bubble is constant. Damping of the bubble dynamics is governed by viscous damping of the surrounding liquid and by acoustic radiation damping, through sound radiated away from the bubble [11, 12, 15, 17, 20, 21, 34, 37]. For the sake of simplicity thermal damping is not included here. Finally, the density of the liquid is large compared to the gas density.

7.2.1 Linearized equations

We often use the linearized equations to describe the bubble dynamics at low driving pressures. For small amplitudes of oscillation the time-dependent radius R can be written as $R = R_0(1 + x(t))$. Through a linearization of the Rayleigh-Plesset equation around the initial radius R_0 we obtain:

$$\ddot{x} + \omega_0\delta\dot{x} + \omega_0^2 x = F(t), \quad (7.2)$$

with x its relative radial excursion, $f_0 = \omega_0/2\pi$ the eigenfrequency of the system and δ the dimensionless damping coefficient. $F(t) = F_0 sin(\omega t)$ is the acoustic forcing term. The eigenfrequency of the system follows from (7.1) and (7.2).

$$f_0 = \frac{1}{2\pi}\sqrt{\frac{1}{\rho R_0^2}\left(3\kappa P_0 + (3\kappa - 1)\frac{2\sigma_w}{R_0}\right)}. \quad (7.3)$$

The total damping coefficient (δ) is given by the sum of the individual damping coefficients. The contribution from the sound radiated by the bubble (δ_{rad}) is:

$$\delta_{rad} = \frac{\frac{3\kappa}{\rho c R_0}\left(P_0 + \frac{2\sigma_w}{R_0}\right)}{\omega_0} \approx \frac{\omega_0 R_0}{c}, \quad (7.4)$$

and the viscous contribution (δ_{vis}) is:

$$\delta_{vis} = \frac{4\nu}{\omega_0 R_0^2}.$$ (7.5)

The resonance frequency of the system is then obtained from:

$$f_{res} = f_0 \sqrt{1 - \frac{\delta^2}{2}}.$$ (7.6)

For a free gas bubble the damping coefficient is negligible. The surface tension is negligible in the mm size range and the resonance frequency is given by the Minnaert frequency [28]:

$$f_{res} \approx f_0 = \frac{1}{2\pi} \sqrt{\frac{3\kappa P_0}{\rho R_0^2}}.$$ (7.7)

For an air bubble in water we then recover the common rule of thumb for the bubble resonance $f_0 R_0 \approx 3$ mm kHz. It should be noted that for bubbles with a radius < 10 µm the surface tension cannot be neglected.

Assuming a steady-state response ($t \to \infty$), substitution into equation 7.2 gives the absolute relative amplitude of oscillation:

$$|x_0| = \frac{F_0}{\sqrt{\left(\omega_0^2 - \omega^2\right)^2 + \left(\delta\omega\omega_0\right)^2}}.$$ (7.8)

For small damping, as in the case of a free gas bubble, the amplitude of oscillations of a bubble driven at a frequency well below its resonance frequency is inversely proportional to the effective 'mass' and the eigenfrequency squared of the system (stiffness-controlled). Well above the resonance frequency the amplitude of oscillation is inversely proportional to the effective 'mass' of the system (inertia-driven). Close to resonance the amplitude of oscillation is inversely proportional to the damping coefficient, the eigenfrequency squared and the effective 'mass' of the system [25].

7.3 Coated bubbles

Ultrasound contrast agents are encapsulated with a phospholipid, protein, palmitic acid or polymer coating. The coating shields the water from the gas, reducing the surface tension to prevent the bubbles from dissolution. Several Rayleigh-Plesset type models have been derived for coated bubbles. Church [6] derived a theoretical model for a coated bubble assuming that the gas core is separated from the liquid by a layer of an incompressible, solid elastic material. The shell has a finite thickness and the shell elasticity and the shell viscosity depend on the rigidity of the shell and the thickness of the shell. Commercial 1st generation Albunex (Mallinckrodt) microbubbles have an albumin shell and remain stable for an extended period of time at atmospheric pressure. Therefore, in Church's model it is assumed that the elastic shell counteracts the capillary pressure ($P_{g0} = P_0$) which stabilizes the bubble against dissolution.

The second generation contrast agents have a more flexible phospholipid shell. The commercially available contrast agents Sonovue® (Bracco), Definity (Lantheus Medical Imaging) and Sonazoid (GE) consist of a monolayer of phospholipids with a thickness of a few nanometers. Various models account for a (phospholipid) coating by assuming a viscoelastic thin shell, see for example De Jong et al. [7], Hoff et al. [18] and more recently Sarkar et al. [36]. The Rayleigh-Plesset type models account for the shell by an effective surface tension ($\sigma(R)$) and the addition of a friction term (S_{fric}) due to the shell elasticity and viscosity, respectively.

$$\rho\left(\ddot{R}R + \frac{3}{2}\dot{R}^2\right) = P_{g0}\left(\frac{R_0}{R}\right)^{3\kappa}\left(1 - \frac{3\kappa\dot{R}}{c}\right) - P_0$$
$$- P_a(t) - 4\nu\rho\frac{\dot{R}}{R} - \frac{2\sigma(R)}{R} - 4S_{fric}\frac{\dot{R}}{R^2}. \tag{7.9}$$

Hoff et al. [18] modified Church's model to account for the thin shell by reducing the equation of Church to a form similar to that of equation 7.9. The effective surface tension and the shell viscosity in the various models are given in Table 7.1. The effective surface tension changes as a function of the bubble radius, see Figure 7.3 for a plot for the various shell models. The parameters are chosen to be comparable in the models ($S_p = 2E_s = 6G_s d_{sh0} = 1.1$ N/m) for the shell elasticity and ($S_f = 16\pi\kappa_s = 48\pi\mu_s d_{sh0} = 2.7 \cdot 10^{-7}$ kg/s) for the shell viscosity, as reported by Gorce et al. [16]. In this regime, the slope of the effective surface tension as a function of the bubble radius is similar for the

Table 7.1. Values for the initial gas pressure in the bubble (P_{g0}), the effective surface tension $\sigma(R)$ and the shell viscosity S_{fric} for three elastic shell models. For comparison the values for an uncoated bubble (Rayleigh-Plesset) are also given

Model	$P_{g0}\,[\mathrm{N/m^2}]$	$\sigma(R)\,[\mathrm{N/m}]$	$S_{fric}\,[\mathrm{kg/s}]$
Rayleigh-Plesset	$P_0 + \dfrac{2\sigma_w}{R_0}$	σ_w	–
Church [6], Hoff et al. [18]	P_0	$6G_s d_{sh0}\dfrac{R_0^2}{R^2}\left(1 - \dfrac{R_0}{R}\right)$	$3\mu_s d_{sh0}\dfrac{R_0^2}{R^2}$
De Jong et al. [7]	$P_0 + \dfrac{2\sigma_w}{R_0}$	$\sigma_w + S_p\left(\dfrac{R}{R_0} - 1\right)$	$\dfrac{S_f}{16\pi}$
Sarkar et al. [36]	P_0	$\sigma(R_0) + E_S\left(\dfrac{R^2}{R_E^2} - 1\right)$	κ_s

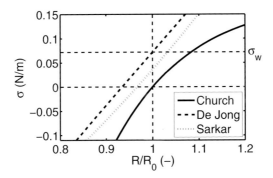

Fig. 7.3. The effective surface tension as a function of the bubble radius ($R_0 = 2 \, \mu m$) for the different models accounting for a purely elastic shell

models by De Jong et al. [7] and Sarkar et al. [36]. The main difference between the models is found for the effective surface tension at the initial bubble radius ($\sigma(R_0)$). It equals σ_w for the model by De Jong et al. [7] and it varies for the model by Sarkar et al. [36]. In this example we choose $\sigma(R_0) = 0.036 \, \text{N/m}$ for the model of Sarkar et al. [36]. The model of Church [6], modified by Hoff et al. [18] for a thin shell, has a lower initial effective surface tension, $\sigma(R_0) = 0 \, \text{N/m}$, and has a different slope. Note that the effective surface tension in these models is not bound and the effective surface tension can become negative and larger than σ_w.

Marmottant et al. [26] introduced a model which seems to be more applicable for high amplitude oscillations. The model accounts for an elastic shell and also for buckling and rupture of the shell. When the shell is elastic, compression of the bubble increases the phospholipid concentration. Therefore, in the elastic regime the effective surface tension decrease is a linear function of the area under compression. Further compression leads to such high phospholipid concentrations that the shell tends to buckle and the effective surface tension vanishes. On the other hand, expansion decreases the phospholipid concentration and leads to rupture. It is assumed that the effective surface tension will relax to σ_w. The effective surface tension using equation 7.9 for the three regimes is given by:

$$\sigma(R) = \begin{cases} 0 & \text{if} \quad R \leq R_{buckling} \\ \chi\left(\dfrac{R^2}{R_{buckling}^2} - 1\right) & \text{if} \quad R_{buckling} \leq R \leq R_{breakup} \\ \sigma_w & \text{if ruptured and } R \geq R_{breakup} \end{cases} \qquad (7.10)$$

with χ the shell elasticity and $R_{buckling}$ and $R_{breakup}$ the buckling and breakup radius, respectively. The effective surface tension as a function of the relative radius for the Marmottant model is shown in Figure 7.4. The shell of a UCA microbubble consists of a monolayer of phospholipids. The initial surface ten-

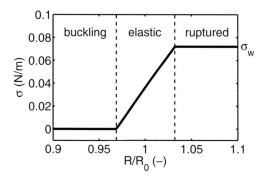

Fig. 7.4. The effective surface tension as a function of the bubble radius ($R_0 = 2\,\mu$m) for the model of Marmottant et al. (2005) including an elastic regime and buckling and rupture of the shell

sion is chosen to be $\sigma(R_0) = 0.036$ N/m similar to the example of the Sarkar model. The choice of $\sigma(R_0)$ in combination with the value for the shell elasticity $\chi = S_p/2 = 0.55$ N/m results in $R_{buckling} = 0.97\,R_0$ and $R_{breakup} = 1.03\,R_0$. In this example the bubble is assumed to rupture when the surface tension reaches σ_w. The shell viscosity in equation 7.9 is given by $S_{fric} = \kappa_s$. As will be shown in the following paragraph, the elasticity of the shell increases the eigenfrequency of the bubble while the shell viscosity increases the total damping of the system.

7.3.1 Linearized equations

The bubble resonance frequency and its corresponding damping coefficient are derived in a similar way as in section 7.2.1. For the model of De Jong et al. [7] the eigenfrequency and the total damping ($\delta_{tot} = \delta_{rad} + \delta_{vis} + \delta_{shell}$) are given by:

$$f_0 = \frac{1}{2\pi}\sqrt{\frac{1}{\rho R_0^2}\left(3\kappa P_0 + (3\kappa - 1)\frac{2\sigma_w}{R_0} + \frac{2S_p}{R_0}\right)} \tag{7.11}$$

$$\delta_{tot} = \frac{\dfrac{3\kappa}{\rho c R_0}\left(P_0 + \dfrac{2\sigma_w}{R_0}\right)}{\omega_0} + \frac{4\nu}{\omega_0 R_0^2} + \frac{S_f}{4\pi\rho R_0^3\omega_0}. \tag{7.12}$$

The eigenfrequency of a coated bubble has two contributions: one part that is similar to the eigenfrequency of an uncoated bubble and an elastic shell contribution. The shell viscosity S_f increases the damping for a coated bubble. Figure 7.5 shows the eigenfrequency and resonance frequency for an uncoated and coated bubble. The resonance frequency and the eigenfrequency of the uncoated bubble agree to within graphical resolution. The eigenfrequency of a coated bubble in comparison to an uncoated microbubble is higher due to

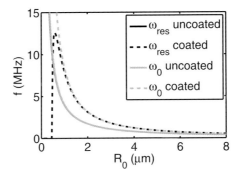

Fig. 7.5. The resonance frequency as a function of the initial bubble radius (R_0) for an uncoated (dotted black line) and coated microbubble (dashed black line). For comparison the eigenfrequency is plotted (grey). The resonance frequency and the eigenfrequency of the uncoated bubble agree to within graphical resolution

the shell elasticity. The damping has a negligible influence on the resonance frequency for an uncoated bubble and for coated bubbles with $R_0 > 1$ µm. Figure 7.6 shows the resonance curve of an uncoated and coated microbubble with a resting radius of 2 µm. The amplitude of oscillation and the resonance frequency are normalized to the maximum amplitude of oscillation and the resonance frequency of the uncoated bubble, respectively. Both the damping and eigenfrequency increase for a coated microbubble, while the effective 'mass' stays the same. The amplitude of oscillation at resonance is therefore lower when the bubble has a shell, see Figure 7.6. Below resonance, neglecting the influence of the damping, the system is stiffness driven. The shell increases the stiffness of the system and the amplitude of oscillation below resonance is therefore lower for a coated microbubble. Far above resonance the amplitude

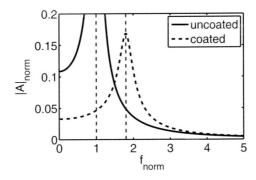

Fig. 7.6. The amplitude of oscillation for an uncoated and coated bubble with $R_0 = 2$ µm, normalized with the maximum amplitude of oscillation of the uncoated microbubble. The driving frequency is normalized to the resonance frequency of the uncoated microbubble

of oscillation is inversely proportional to the effective 'mass' of the system. Consequently above resonance the amplitude of oscillation does not depend on the shell properties.

7.3.2 Optical and acoustical characterization

The theoretical models are validated through experiments on single bubbles. Acoustical and optical experiments reveal the response of UCA microbubbles and both have there own particular advantages and disadvantages. In acoustical experiments, the scattered pressure or pressure-time P-t curve is recorded. Acoustic characterization is relatively inexpensive and has the advantage of a high sampling rate using long pulse sequences. The scattered pressure of a single bubble however is limited (order 1 Pa) and close to the noise level of our detection system. The transducer focus is in the order of the acoustic wavelength and a bubble must be isolated in the in vitro setup. In optical experiments high-speed cameras are used to record the radial response or radius-time R-t curve of single bubbles. Such a camera should be able to temporally resolve the dynamics of the microbubbles which is driven at MHz frequencies. The required framerates makes the construction of such a camera expensive. The recording time is limited by the number of frames. Optical characterization of single microbubbles is relatively easy. The Brandaris 128 camera, see Figure 7.2, was especially designed for this purpose [5]. The camera uses a fast rotating mirror (max 20,000 rps) to sweep the image across 128 highly sensitive CCDs. At maximum speed an interframe time of 40 ns, which corresponds to a framerate of 25 Mfps is obtained. Figure 7.7 shows a sequence of 25 frames recorded with the Brandaris 128 camera at a framerate of 13.5 Mfps. The applied acoustic pulse has a frequency of 2.7 MHz and a pressure of 30 kPa. The accompanying R-t curve of the microbubble from the Brandaris movie is shown in Figure 7.8. The maximum amplitude of oscillation is 200 nm corresponding to a relative amplitude of 10%.

The first fitting of shell parameters for SonoVue® were performed acoustically on a microbubble suspension [16]. Recently, optical R-t curves of single UCA microbubbles (SonoVue®) were recorded and fitted, to an elastic shell model (Hoff's model), by Chetty et al. [4]. In Chetty et al. [4] the shell thickness and shell viscosity was fixed and it was found that the shell elasticity increases with increasing bubble radius. The experiments were performed with a single applied frequency (0.5 MHz) and pressure (40-80 kPa). No experiments were performed to test the validity of the shell parameters for the same bubble at different ultrasound frequencies and pressures. Van der Meer et al. [38] insonified single UCA microbubbles (BR-14) consecutively with 11 ultrasound pulses, increasing the frequency for each pulse, near resonance. A resonance curve was then obtained by plotting the amplitude of oscillation as a function of the applied frequency. A fit of the resonance curve to the linearized shell model of Marmottant et al. [26] then resulted in the shell elasticity and shell viscosity. In contrast to Chetty et al. [4], Van der Meer

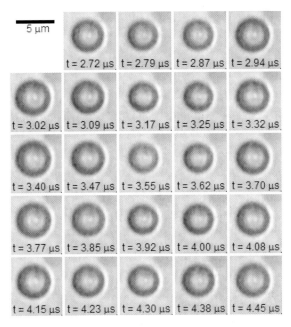

Fig. 7.7. Sequence of 24 frames of a 2.2 μm radius bubble recorded with the Brandaris 128 camera at a framerate of 13.5 Mfps

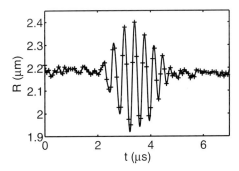

Fig. 7.8. The R-t curve of the same bubble as in Figure 7.7. The bubble is insonified with an ultrasound pulse with a frequency of 2.7 MHz and an acoustic pressure of 30 kPa

et al. [38] found that the shell elasticity was nearly constant while the shell viscosity decreases with decreasing dilatation rate $\left(\dot{R}/R\right)$. One should note that all the above experiments were performed at or in close proximity to a (capillary) wall. In the following section we will discuss the influence of a rigid wall on bubble dynamics.

De Jong et al. [8] reported on an observation of coated microbubbles at low applied acoustic pressures, where the bubbles compress, but hardly expand.

This highly non-linear effect, referred to as compression-only behavior, occurs for 40% of the bubbles even at pressures as low as 50 kPa. Remarkably all bubbles with an initial radius less than 2 μm show compression only behavior at a frequency of 1 MHz. Compression only behavior is not observed for uncoated bubbles and cannot be described by a model accounting for purely an elastic shell. Actually, the purely elastic shell models predict even a decrease of the non-linear behavior of the coated microbubbles as compared to the dynamics of an uncoated microbubble. The model of Marmottant et al. [26] accounting for an elastic shell and for buckling and rupture of the shell predicts compression-only behavior. As stated by Marmottant et al. [26] the compression modulus in the elastic state is much higher than in the buckled or ruptured state. For a bubble where the resting radius is the buckling radius it is much harder to expand than to compress and results in compression only behavior of the bubble [26].

Emmer et al. [10] showed an oscillation threshold for coated microbubbles with a radius smaller than 2.5 μm at a driving frequency of 1.7 MHz. Below a certain pressure (30–120 kPa) the bubbles hardly oscillate while above this threshold the amplitude of oscillation increases linear with the applied acoustic pressure. This non-linear effect occurs at low acoustic pressures and is not yet fully understood. The non-linear behavior of the UCA microbubbles can be exploited to improve present contrast enhanced imaging techniques.

7.4 Bubble dynamics near a rigid wall

In this section we discuss the influence of a rigid wall on the bubble dynamics. We start with the simplest approach, the so-called method of images, to simulate the influence of a wall. We also study the system experimentally by recording the radial bubble dynamics at the wall and in free space, away from the wall.

7.4.1 Method of images

In the literature, several extensions to the bubble dynamics equations have been made to account for the presence of a rigid wall. All the models described here are based on the method of images depicted in Figure 7.9. If the wall is rigid, the specific acoustic impedance $Z = \rho c$ is infinite, and no energy crosses the wall. To describe the acoustic (or equivalently the fluid-mechanical field) the wall is replaced by an identical image bubble oscillating in-phase with the real bubble and positioned at the mirrored image point. The real bubble dynamics is influenced by the pressure emitted by the image bubble. The dynamics of a coated bubble near a rigid wall is therefore described by a Rayleigh-Plesset type equation including the radiated pressure of the image

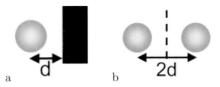

Fig. 7.9. The method of images. In (**a**) the actual situation, where the bubble is located at a distance d from the rigid wall, (**b**) shows the method of images in which the wall is replaced by an image bubble

bubble:

$$\rho\left(\ddot{R}R + \frac{3}{2}\dot{R}^2\right) = P_{g0}\left(\frac{R_0}{R}\right)^{3\kappa}\left(1 - \frac{3\kappa\dot{R}}{c}\right) - P_0$$

$$- P_a(t) - 4\nu\rho\frac{\dot{R}}{R} - 4S_{fric}\frac{\dot{R}}{R^2} - \frac{2\sigma(R)}{R} - \rho\frac{\partial}{\partial t}\left(\frac{\dot{R}R^2}{2d}\right),$$

(7.13)

where d represents the distance between the bubble and the wall. For a bubble positioned directly at the wall, such as bubbles floating up against the capillary wall, the distance d is simply given by the bubble radius $R(t)$. In this case the bubble dynamics equation becomes:

$$\rho\left(\frac{3}{2}\ddot{R}R + 2\dot{R}^2\right) = P_{g0}\left(\frac{R_0}{R}\right)^{3\kappa}\left(1 - \frac{3\kappa\dot{R}}{c}\right) - P_0$$

$$- P_a(t) - 4\nu\rho\frac{\dot{R}}{R} - 4S_{fric}\frac{\dot{R}}{R^2} - \frac{2\sigma(R)}{R}.$$

(7.14)

Note that all assumptions made previously for the Rayleigh-Plesset equation for an uncoated microbubble remain valid. Therefore the bubble must remain spherical, which may not be strictly true in the experimental situation. For example we know that bubbles deform close to the wall at pressures of 140 kPa [39].

7.4.2 Linearized equations

The rigid wall increases the effective 'mass' of the bubble $3/2$ times (increased prefactor of the inertial term) resulting in a decrease of the eigenfrequency and damping. For an (un)coated bubble at a wall, the eigenfrequency and damping are derived in a similar way as in section 7.2.1. The eigenfrequency and damping for an uncoated bubble reduce to:

$$f_0^{wall} = \sqrt{\frac{2}{3}}f_0^{free} \approx 0.8f_0^{free}$$

$$\delta^{wall} = \sqrt{\frac{2}{3}}\delta^{free} \approx 0.8\delta^{free}.$$

(7.15)

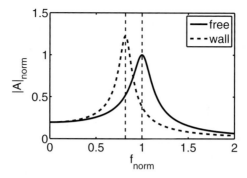

Fig. 7.10. Resonance curves for an uncoated bubble with a initial radius of 2 μm in free space (solid), at a rigid wall. The frequency and the amplitude are normalized with the resonance frequency and amplitude of oscillation at resonance in the free case

Figure 7.10 shows the resonance curve of a coated bubble in free space (solid) and at the wall (dashed). The amplitude of oscillation and the applied frequency are normalized to that of the bubble in free space. The amplitude of oscillation at resonance is $\sqrt{3/2}$ larger for a bubble at a wall than for a free bubble, see Figure 7.10. Below the resonance frequency the amplitude of oscillations remains unchanged (stiffness-controlled) while above resonance the amplitude of oscillation is 1.5 times smaller for a bubble at a wall because of the increased effective 'mass' of the system.

7.4.3 Observations

Bubbles injected in an in vitro setup (e.g. capillary or flow cell) float up due to buoyancy until they reach the top wall. Due to the limited focal depth of the microscope objective the rising bubbles are difficult to capture in free space. The radial bubble dynamics is therefore traditionally captured with the bubbles positioned against the top wall of the capillary. In order to study the influence of a wall on the bubble dynamics, the radial bubble response should also be recorded in free space. To investigate the bubbles in free space we trap the microbubbles and control their position in 3D space.

The pioneering work on particle trapping was that of Ashkin [1], who showed the first optical trap for high-index particles $\left(n = n_{particle}/\,n_{medium} > 1\right)$. Optical trapping has undergone rapid development, attracting special attention from the life science discipline, as nanoparticles, cells, viruses and bacteria can be micromanipulated and controlled. A high-index particle is trapped in the intensity maximum of the laser beam focus. The optical gradient force is directed towards the intensity maximum, while the scattering force, away from the intensity maximum cancels out. Low-index particles, such as the microbubbles studied here $\left(n = n_{particle}/n_{medium} < 1\right)$, are repelled by both the scattering force and the gradient force. They can, however, be trapped in the intensity minimum in the center of a donut-shaped Laguerre-Gaussian (LG) laser beam [13]. LG beams are obtained by modulating the phase of a Gaussian beam, using diffractive optical elements (DOEs).

To investigate the influence of a wall on the dynamics of coated microbubbles, Garbin et al. [14] combined an optical tweezers setup with the ultra-high speed Brandaris 128 camera. This enabled the micromanipulation of single UCA microbubbles in 3D space, thereby temporally resolving bubble dynamics, which as a result allowed for a detailed investigation of the influence of the flow cell wall. Figure 7.11 shows the R-t curves of a single bubble with a radius of 2.5 μm insonified with a frequency of 2.25 MHz and a pressure of 200 kPa. The top figure shows the R-t curve of the bubble at the flow cell wall, while the middle figure shows the R-t curve of the very same bubble positioned 50 μm away from the wall in free space. The bottom figure shows the bubble response back at the wall and the amplitude of oscillation is similar as in the first experiment where the bubble was also located at the wall. Garbin et al. [14] reported a decrease in the amplitude of oscillations at the flow cell wall by over 50%.

The following three possible explanations for the decrease in amplitude at the wall were discussed in Garbin et al. [14]. First, a change in the resonance frequency caused by the presence of the wall. This explanation is consistent with the models described here, that show a decrease of the resonance frequency by 20%. A viscous boundary layer formed near the wall inducing shear stresses and additional dissipation was suggested as a second mechanism to explain the observed behavior. Up to now simulations describing the boundary layer approach have failed to predict the decrease of the amplitude of oscilla-

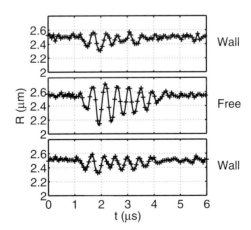

Fig. 7.11. R-t curves for a single microbubble with a radius of 2.5 μm. The applied ultrasound has a frequency of 2.25 MHz and a pressure of 200 kPa. The microbubble is insonified at the wall (top), in free space 50 μm away from the wall (middle) and again at the wall (bottom)

tions near the wall. A third explanation that was given in Garbin et al. [14] were non-spherical oscillations caused by the non-symmetrical interaction of the oscillating bubble with the wall. In the setup by Garbin et al. [14] non-spherical oscillations could not be observed due to the optical configuration employed. The optical axis was perpendicular to the flow cell wall, i.e. it was always observed in top-view. Voss et al. [39] showed, with a setup allowing both, a side-view (optical axis parallel to the wall) and a top-view that the oscillations of UCA microbubbles may appear spherical in top-view and can be quite asymmetric in side-view.

Until now dissipation caused by translational oscillations due to secondary Bjerknes force was not considered. A microbubble in an incident acoustic wave will experience an acoustic radiation force. The force $\mathbf{F}$ depends on the volume of the bubble V and the acoustic pressure gradient $\boldsymbol{\nabla} P$:

$$\mathbf{F} = -V\boldsymbol{\nabla} P. \tag{7.16}$$

The force is called the (primary) Bjerknes force [2] if it is caused by the incident acoustic wave. An oscillating microbubble radiates sound (pressure waves) and thereby induces an acoustic radiation force on its neighboring bubble, termed the secondary Bjerknes force. The Bjerknes force leads to an alternating attractive and repulsive translation. Bubbles oscillating in phase have a net attractive force while bubbles oscillating out of phase repel each other. Consequently, a microbubble near a rigid wall will experience a net attractive force towards the wall, as its image bubble oscillates in phase.

For the simulations described here the wall is considered as an infinitely thick rigid wall. No energy passes the wall and the ultrasound will be fully reflected at the wall. In our experiments the wall is acoustically transparent to allow ultrasound to enter the flow cell and to prevent unwanted reflections. For such a compliant wall the (complex) amplitude of the image bubble needs to be adapted to the wall properties and this will be the subject of a future study.

7.5 Conclusions

A phospholipid coating is accounted for by a thin viscoelastic layer. The shell elasticity increases the eigenfrequency while a shell viscosity increases the total damping and decreases the amplitude of oscillations of coated bubbles. Most models, valid for low amplitude oscillations, describe the shell as fully elastic [7,18,36] and predict a decrease in the linear and the non-linear behavior of coated microbubbles. Experiments show non-linear behavior for coated microbubbles at low applied acoustic pressure, for example, compression only behavior and thresholding behavior. The model of Marmottant et al. [26] including buckling and rupture of the shell is valid for high amplitude oscillations and predicts the observed compression only behavior.

With the method of images a decrease in the resonance frequency and an increase of the amplitude at resonance is predicted for uncoated and coated bubbles. These models do not include boundary layers and the oscillations are assumed to remain spherical at all times.

The change in the dynamics of UCA microbubbles near a wall Garbin et al. [14] is important for molecular imaging applications. The next step after understanding the influence of a phospholipid shell and a wall is to understand the influence of ligands on the coating, and targeting the wall. Optimization of pulse-echo techniques can be done when there is a full understanding of the complex systems.

Acknowledgements

We thank many of our collaborators on this subject for their great help: Benjamin Dollet, Marcia Emmer, Valeria Garbin, Todd Hay, Sascha Hilgenfeldt, Detlef Lohse, Philippe Marmottant, Sander van der Meer, Jeroen Sijl, and Timo Rozendal. We gratefully acknowledge the technical assistance of Wim van Alphen, Leo Bekkering, Martin Bos, Gert-Wim Bruggert, Jan Honkoop, Frits Mastik, Cees Pakvis, and Geert Springeling in constructing and improving our experimental setups. We acknowledge Bracco Research S.A. (Geneva) for providing BR-14 vials and for stimulating discussions. This work is financially supported by TAMIRUT, under project nr. NMP4-CT-2005-016382.

References

1. Ashkin A (1986) Observation of a single-beam gradient force optical trap for dielectric particles. Optics Letters 11(5):288
2. Bjerknes V (1906) Fields of Force. Columbia University Press
3. Brock-Fisher G, Poland M and Rafter P (1996) Means for increasing sensitivity in non-linear ultrasound imaging systems US patent no 55775
4. Chetty K, Stride E, Sennoga C, Hajnal J and Eckersley R (2008) High-speed optical observations and simulation results of sonovue microbubbles at low-pressure insonation. IEEE Transactions on Ultrasonics, Ferroelectrics, and Frequency Control 55(6):1333–1342
5. Chin C, Lancee C, Borsboom J, Mastik F, Frijlink M, de Jong N, Versluis M and Lohse D (2003) Brandaris 128: A digital 25 million frames per second camera with 128 highly sensitive frames. Review of Scientific Instruments 74:5026–5034
6. Church C (1995) The effects of an elastic solid surface layer on the radial pulsations of gas bubbles. The Journal of the Acoustical Society of America 97(3):1510–1521
7. De Jong N, Cornet R and Lancee CT (1994) Higher harmonics of vibrating gas-filled microspheres. part one: simulations. Ultrasonics 32:447–453
8. De Jong N, Emmer M, Chin C, Bouakaz A, Mastik F, Lohse D and Versluis M (2007) "compression-only" behavior of a phosphorlipid-coated contrast bubbles. Ultrasound in Medicine and Biology 33(4)
9. Doinikov A (2001) Translational motion of two interacting bubbles in a strong acoustic field. Physical Review E 64(2):026,301

10. Emmer M, Wamel AV, Goertz D and Jong ND (2007) The onset of microbubble vibration. Ultrasound in Medicine and Biology 33(6):941–949
11. Flynn H (1975a) Cavitation dynamics. i. a mathematical formulation. The Journal of the Acoustical Society of America 57(6):1379–1396
12. Flynn H (1975b) Cavitation dynamics: Ii. free pulsations and models for cavitation bubbles. The Journal of the Acoustical Society of America 58(6):1160–1170
13. Gahagan K (1996) Optical vortex trapping of particles. Optics Letters 21(827):11
14. Garbin V, Cojoc D, Ferrari E, Fabrizio ED, Overvelde M, van der Meer S, de Jong N, Lohse D and Versluis M (2007) Changes in microbubble dynamics near a boundary revealed by combined optical micromanipulation and high-speed imaging. Applied Physics Letters 90:114,103
15. Gilmore F (1952) The growth or collapse of a spherical bubble in a viscous compressible liquid. Tech. rep., Hydrodynamics Laboratory, California Institute Technology, Pasadena, report 26–4
16. Gorce JM, Arditi M and Schneider M (2000) Influence of bubble size distribution on the echogenicity of ultrasound contrast agents. Investigative Radiology 35(11):661–671
17. Herring C (1941) Theory of the pulsations of the gas bubble produced by an underwater explosion. Tech. rep., OSRD report 236
18. Hoff L, Sontum P and Hovem J (2000) Oscillations of polymeric microbubbles: Effect of the encapsulating shell. The Journal of the Acoustical Society of America 107(4):2272–2280
19. Hope Simpson D, Ting CC and Burns P (1999) Pulse inversion doppler: a new method for detecting nonlinear echoes from microbubble contrast agents. IEEE Transactions on Ultrasonics, Ferroelectrics, and Frequency Control 46(2):372–382
20. Keller J and Kolodner I (1956) Damping of underwater explosion bubble oscillations. Journal of Applied Physics 27(10):1152–1161
21. Keller JB and Miksis M (1980) Bubble oscillations of large amplitude. The Journal of the Acoustical Society of America 68:628–633
22. Klibanov A (2002) Ultrasound Contrast Agents: Development of the Field and Current Status, Topics in Current Chemistry 222
23. Lankford M, Behm C, Yeh J, Klibanov A, Robinson P and Linder J (2006) Effect of microbubble ligation to cells on ultrasound signal enhancement: implications for targeted imaging. Investigative Radiology 41(10)
24. Lauterborn W (1976) Numerical investigation of nonlinear oscillations of gas bubbles in liquids. The Journal of the Acoustical Society of America 59(2):283–293
25. Leighton T (1994) The Acoustic Bubble. Academic Press Inc. San Diego
26. Marmottant P, van der Meer S, Emmer M, Versluis M, de Jong N, Hilgenfeldt S and Lohse D (2005) A model for large amplitude oscillations of coated bubbles accounting for buckling and rupture. The Journal of the Acoustical Society of America 118:3499–3505
27. Marmottant P, Versluis M, Jong ND, Hilgenfeldt S and Lohse D (2006) High-speed imaging of an ultrasound-driven bubble in contact with a wall: "narcissus" effect and resolved acoustic streaming. Experiments in Fluids 41(2):147–153
28. Minnaert M (1933) On musical air-bubbles and sounds of running water. Philosophical Magazine 16:235–248

29. Neppiras E and Noltingk B (1951) Cavitation produced by ultrasonics: Theoretical conditions for the onset of cavitation. Proceedings of the Physical Society Section B 64(12):1032–1038

30. Noltingk B and Neppiras E (1950) Cavitation produced by ultrasonics. Proceedings of the Physical Society Section B 63(9):674–685

31. Plesset M (1949) The dynamics of cavitation bubbles. Journal of Applied Mechanics 16:277–282

32. Poritsky H (1952) The collapse or growth of a spherical bubble or cavity in a viscous fluid. Proceedings of the first US National Congress on Applied Mechanics pp 813–821

33. Prosperetti A (1975) Nonlinear oscillations of gas bubbles in liquids. transient solutions and the connection between subharmonic signal and cavitation. The Journal of the Acoustical Society of America 57(4):810–821

34. Prosperetti A, Crum L and Commander K (1988) Nonlinear bubble dynamics. The Journal of the Acoustical Society of America 83(2):502–514

35. Rayleigh L (1917) On the pressure developed in a liquid during the collapse of a spherical cavity. Philosophical Magazine 34:94–98

36. Sarkar K, Shi W, Chatterjee D and Forsberg F (2005) Characterization of ultrasound contrast microbubbles using in vitro experiments and viscous and viscoelastic interface models for encapsulation. The Journal of the Acoustical Society of America 118(1):539–550

37. Trilling L (1952) The collapse and rebound of a gas bubble. Journal of Applied Physics 23(1):14–17

38. Van der Meer S, Dollet B, Chin CT, Bouakaz A, Voormolen M, de Jong N, Versluis M and Lohse D (2007) Microbubble spectroscopy of ultrasound contrast agents. The Journal of the Acoustical Society of America 120:3327–3327

39. Vos H, Dollet B, Bosch J, Versluis M and de Jong N (2008) Nonspherical vibrations of microbubbles in contact with a wall – a pilot study at low mechanical index. Ultrasound in Medicine and Biology 34(4):685–688

40. Zhao S, Ferrara K and Dayton P (2005) Asymmetric oscillation of adherent targeted ultrasound contrast agents. Applied Physics Letters 87(13)

41. Zhao S, Kruse D, Ferrara K and Dayton P (2006) Acoustic response from adherent targeted contrast agents. The Journal of the Acoustical Society of America 120(6):EL63–EL69

Chapter 8

Characterization of Acoustic Properties of PVA-Shelled Ultrasound Contrast Agents

Dmitry Grishenkov, Claudio Pecorari, Torkel B. Brismar, and Gaio Paradossi

Abstract. This work examines the acoustic behavior of ultrasound contrast agents made of poly (vinyl alcohol) (PVA) shelled microbubbles manufactured at three different pH and temperature conditions. Backscattering amplitude, attenuation coefficient and phase velocity of ultrasonic waves propagating through suspensions of PVA contrast agents were measured at temperature values ranging from 24°C to 37°C in a frequency range from 3 MHz to 13 MHz. A significant enhancement of the backscattering amplitude and a weak dependence on temperature were observed. Attenuation and phase velocity, on the other hand, showed higher sensitivity to temperature variations. The dependence on system parameters such as the number of cycles, frequency, and exposure of the peak negative pressure, P_{thr}, at which ultrasound contrast agents fracture was also investigated. The effects of temperature, blood, and, whenever data were available, of the dimension of the microbubbles on P_{thr} are also considered. The large shell thickness notwithstanding, the results of this investigation show that at room temperature, PVA contrast agents fracture at negative peak pressure values within the recommended safety limit. Furthermore, P_{thr} decreases with increasing temperature, radius of the microbubbles, and number of cycles of the incident wave. In conclusion, these results suggest that PVA-shelled microbubbles may offer a potentially viable system to be employed for both imaging and therapeutic purposes.

8.1 Introduction

The first generation of ultrasound contrast agents (UCAs) was made of free air bubbles. These were however quickly dissolved in blood and it was difficult to make the bubbles homogenous in size. They were therefore of little practical use. The currently used UCAs have a thin solid shell made of either proteins (Albumex and Optison), lipids (SonoVue, Definity, and Imagent), or surfactants (ST68). Instead, airless soluble gases such as octafluoropropane, perfluorohexane and sulfurhexafluoride are used. The next generation of UCAs will combine imaging with therapeutic functionalities. By attaching ligands to the surface of the bubbles they can be designed to adhere to the membrane

Paradossi, G., Pellegretti, P., Trucco, A. (Eds.)
Ultrasound contrast agents. Targeting and processing methods for theranostics
© Springer-Verlag Italia, 2010

of diseased cells. This will make it possible to visualize the area of the disease and it will increase the concentration of the bubbles in that area. If the bubbles are loaded with drugs these can be released by increasing the intensity of the focused ultrasound. The local release will make it possible to reach higher local concentrations of toxic drugs compared to if the drugs had been delivered systemically.

In this book the properties of three sets of PVA based microbubbles that have the potential to be used as third generation UCAs are described. This chapter focuses on the fracture properties of the PVA shells of the microbubbles when they are exposed to a focused beam of high-amplitude ultrasonic waves. In particular, the peak negative pressure, P_{thr}, at which PVA shells fracture is investigated as a function of the number of cycles, N, frequency, f (or $\omega = 2\pi g$), and pulse repetition frequency (PRF) of the incident pulse, as well as of temperature, T, of the host medium. The tests have been carried out with both water and blood as surrounding media.

8.2 Materials and methods

8.2.1 Particle size distribution

Three polymer-shelled UCAs labeled MB-pH5-RT, MB-pH5-Cool, and MB-pH2-Cool were examined in this study. They were synthesized from an aqueous solution of PVA modified according to the protocols described by Cavalieri et al. [1]. These PVA solutions were characterized by pH values equal to 2 at 4°C (cool) and 5 at room temperature (RT). The average external diameter and shell thickness of these microbubbles are reported in Table 8.1. Note, in all three cases the ratio between the shell thickness and external diameter which is greater than 30 percent. Microbubble suspensions with concentration values of 3.7×10^5 ml^{-1} (MB-pH5-RT), 3.0×10^6 ml^{-1} (MB-pH5-Cool), and 6.5×10^6 ml^{-1} (MB-pH2-cool) were used.

8.2.2 Experimental setups

To measure the power of signals backscattered by a suspension of microbubbles the setup which is described next was used. A focused transducer (Panametrics

Table 8.1. Average diameter and shell thickness of the microbubbles

Microbubble Type	Average Diameter (μm)	Average Thickness (μm)
MB-pH5-RT	4.1 ± 0.7	0.7 ± 0.1
MB-pH5-Cool	2.7 ± 0.6	0.5 ± 0.1
MB-pH2-Cool	2.6 ± 0.5	0.5 ± 0.1

V309, Waltham, MA, USA) with diameter $d = 12.7$ mm, focal length $\ell = 50$ mm, and central frequency $f = 5$ MHz was employed in a pitch-catch mode. The suspension under examination was confined within a custom-made cell with two acoustic windows formed by thin plastic sheets. The dimension of the cell along the direction of propagation of the inspecting beam was L = 1.4 cm. The distance between the cell and the transducer was chosen so that the beam was focused just behind the front wall of the cell. The transducer was excited by a pulser-receiver (Panametrics PR 5072, Waltham, MA, USA) with broadband pulses and a pulse repetition frequency of 1 kHz. The signals detected by the focused transducer were digitized by an oscilloscope (Tektronix TDS 5052, Tektronix Inc., Beaverton, OR, USA). The amplitude spectra of the time domain signals were averaged over ten samples to decrease the noise level by about 10 dB. The transducer and the cell were placed in a tank containing deionized and degassed water.

Measurements of the attenuation and phase velocity was carried out using a pitch-catch configuration with a flat transducer with nominal frequency $f = 10$ MHz and diameter $d = 12.7$ mm (Panametrics V311). An aluminum block, which was positioned 1 cm after the sample container, was employed as a reflector. The temperature of the whole system was controlled using a heating immersion circulator (Julabo ED, Julabo Labortechnik GmbH, Germany). Measurements were carried out at 24°C, 28°C, 31°C, 34°C, and 37°C.

Finally, in the experimental setup, to measure the peak negative pressure, P_{thr}, the emitting and receiving transducers were arranged so that their focal regions intersected each other within a (nearly) cylindrical container defining the volume occupied by the suspension of UCAs. Two focused transducers operating at a nominal frequency of 1.1 MHz (Precision Acoustics, diameter = 23 mm, focal length = 45 mm) and 2.2 MHz (Krautkramer, Gamma Series, diameter $d = 12.7$ mm, focal length = 50 mm), respectively, were used as emitters. The transducer working at 2.2 MHz was also used as a receiver of the signals generated by the 1.1 MHz emitter, while a third transducer with nominal frequency of 5 MHz (Panametrics, V309, diameter = 12.7 mm, focal length = 50 mm, focal length = 50 mm) was employed as a receiver when the 2.2 MHz transmitter was used. The transducers' axes were orthogonal to the axis of the cylindrical container confining the UCA suspension.

The electrical signal driving the transmitting transducer was generated by a high-power, tone-burst pulser-receiver (SNAP Mark IV, Ritec Inc.), which controls the number of cycles and the PRF of each pulse. Two attenuators were used to vary the intensity of the electrical signal exciting the transducer. The time-domain signals detected by the receiver were digitized at a sampling frequency of 80 MHz by a 14 bit waveform Digitizer (CompuScope 14200, Gage Applied Technologies).

In this study, blood donated for medical blood transfusions was used in accordance with the ethical guidelines of the institution. For transfusion medical reasons this blood is free from thrombocytes and leukocytes. The erythrocyte volume fraction is 60 %. 100 mL of the transfusion blood contains 20 g hemoglobin, 4-8 % plasma, 307 mg NaCl, 315 mg Dextrose, 6 mg adenine and 184 mg mannitol.

Finally, in order to recover estimates of the peak negative pressure, P_{thr}, the time-domain signals were processed according to the method presented by Pecorari and Grishenkov [7]. In this way, the loss of coherence between two consecutive acoustic responses of the suspension is attributed to the occurrence of shell fractures and consequent release of free gas microbubbles into the suspension.

8.3 Results

8.3.1 Backscattered power spectra

Measurements of the backscattered power spectra for the MB-pH5-Cool microbubbles are shown in Figure 8.1. Negligible changes of backscattering were observed with increasing temperature. However, an increase of more than 3 dB is recorded when the temperature varies from 31°C to 37°C. This fact can be considered as a sign that some change has occurred in the mechanical and/or geometrical properties of the microbubbles. Similar observations were recorded on the power of the signals backscattered by the other microbubbles.

Of relevance for imaging applications, the signal produced by the suspension of microbubbles MB-pH5-Cool with concentration values of 3.0×10^6 ml^{-1} is more than 20 dB above the noise. By the term "noise" we consider the backscattered spectrum acquired when no microbubbles but only distilled and deionized water was present within the cell. This spectrum was recorded with the same procedure and equals approximately 2 dB.

A comparison with similar results reported by other authors [5, 6, 8]) investigating polymer-shelled contrast agents would indicate that PVA UCAs require a concentration (measured in #-particles/ml) ten times lower that those examined by Lathia et al. [5], and one hundred times lower that those used by Lavisse et al. [6].

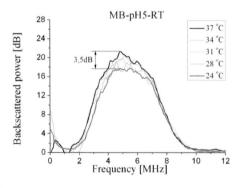

Fig. 8.1. Average backscattered power from a suspension of microbubbles MB-pH5-RT

On the contrary, Wheatley et al. [8] report backscattered enhancement of the same order of magnitude employing UCA concentrations at least three times lower than those used in our work.

8.3.2 Attenuation and dispersion

Figure 8.2a–c shows the frequency dependence of the attenuation coefficient of the three types of microbubbles introduced earlier. The monotonic increase of the attenuation with frequency can be observed at all temperatures considered in this test. The measurements illustrate also the rise of the attenuation coefficient as the temperature increases with variations of the order of 30 percent above 10 MHz. At values of the concentration leading to the same volume fraction of microbubbles, the suspension of MB-pH5-Cool microbubbles produced the largest wave attenuation reaching a value close to 15 dB/cm at 13 MHz, 37°C and at a concentration of 3.0×10^6 ml^{-1}.

The frequency dependence of the phase velocity at temperature values between 24°C and 37°C was recovered using the same time-domain signals employed to obtain data regarding the attenuation coefficient. In Figure 8.2d–f, the straight lines represent the values of the phase velocity in pure water at the corresponding temperature. The monotonic increase of the phase velocity with increasing frequency is typical of dispersion phenomena in spectral

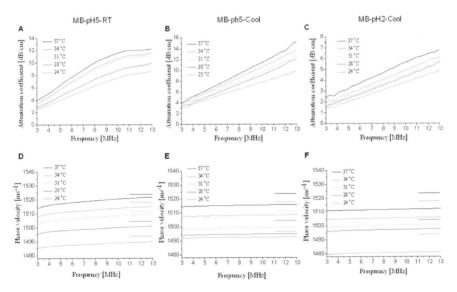

Fig. 8.2. Attenuation coefficient and phase velocity vs. frequency at temperature varying between 24°C and 37°C: (**a**) and (**d**) MB-pH5-RT; (**b**) and (**e**) MB-pH5-COOL; (**c**) and (**f**) MB-pH2-COOL. The horizontal lines in a phase velocity plot indicate the value of the phase velocity in pure water at the corresponding temperature

regions far from resonances. The distance from the closest resonance of the frequency region explored in these measurements is also confirmed by the modest frequency dependence of the phase velocity.

In these experiments, the suspension of microbubbles labeled MB-pH5-RT produced the largest dispersion in the frequency range between 3 MHz to 13 MHz. Furthermore, at the upper end of this range an ultrasonic wave appears to propagate with a phase velocity which approaches its value in pure water more closely in this suspension than in those of MB-pH5-Cool and MB-pH2-Cool contrast agents. The larger dimensions and polydispersity of the MB-pH5-RT microbubbles might partially explain these observations.

The data acquired at 24°C can be used also to extract information concerning viscoelastic and structural properties of the PVA microbubbles of interest. To this end, Church's model [3] must be generalized to include the frequency dependence of the dynamic viscoelastic modulus of the polymeric shells [4]. Estimates of the frequency dependence of the storage and loss moduli can be recovered, which suggest some similarity between the microstructure of the PVA shell and that of semi-flexible polymeric networks.

8.3.3 Ultrasound-induced fracture of PVA microbubbles

Figure 8.3a–c reports the values of P_{thr} for the microbubbles labeled MB-pH5-RT, MB-pH5-Cool, and MB-pH2-Cool, respectively, as a function of the number of cycles of the incident beam produced by the focused transducer with central frequency $f = 2.2$ MHz. These data were acquired both at 24°C (blue symbols) as well as at 37°C (red symbols). For these tests, suspensions with concentration values of 3.75×10^5 ml^{-1}, 3×10^6 ml^{-1} and 5.2×10^6 ml^{-1} were used for the microbubbles labeled MB-pH5-RT, MB-pH5-Cool, and MB-pH2-Cool, respectively.

The first important point suggested by these data concerns the nature of the fracture process. Since P_{thr} decreases with increasing number of cycles, N, fatigue would appear to play an important role in fracturing the shell of all three types of microbubbles. In other words, the results of Figure 8.3 show that the original structure of the shells is affected by "defects" which act as stress concentrators facilitating the evolution of damage during cyclic loading. The more severe the shell's defects are prior to cyclic loading, the sooner the final stage of this process, i.e., the macroscopic fracture of the shell is reached. Results showing this trend have previously been reported by Chen et al. [2], who investigated the same dynamic properties of four types of UCAs. Interestingly, their data on thick polymer-shelled labeled BiSpheres-0.7X at 3.5 MHz ($P_{thr} = 0.95$ MPa for $N = 10$) are of the same order of magnitude as those reported in this work. The shell of BiSphere contrast agents is a two layer structure. The internal layer is lactide polymer with a thickness of 226.8 nm and provides stability to the structure.

Secondly, of the three types of microbubbles considered here, the ones synthesized at higher temperature and pH value, which also have the largest

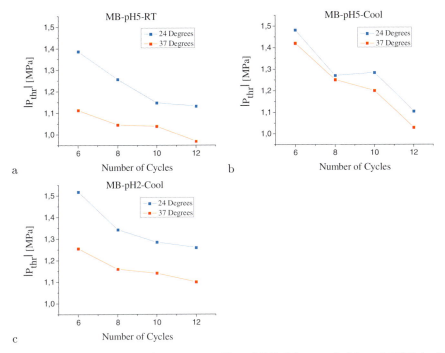

Fig. 8.3. Absolute value of P_{thr} versus N at $24°$C (blue symbols) and $37°$C (red symbols) for (**a**) MB-pH5-RT, (**b**) MB-pH5-Cool, and (**c**) MB-pH2-Cool microbubbles. The frequency of the incident beam was 2.2 MHz

average diameter (see Table 8.1), are the easiest to break. Conversely, those synthesized at lower temperature and pH value, have the lower average diameter and require the highest amount of energy to fracture. Furthermore, the effect of the magnitude of the diameter of the microbubbles on P_{thr} is illustrated by the comparison between the value of P_{thr} at $24°$C and N = 6 of the microbubbles UCAs MB-pH5-Cool (1.4 MPa) and that reported by Pecorari and Grishenkov [7] for the same type of UCAs (0.8 MPa). The average diameter of the microbubbles used in this work is equal to 2.6 µm, while it was equal to 5 µm in the previous investigation. The mechanisms at molecular level which are behind this host of phenomena are still a matter of conjecture.

The dependence of P_{thr} on the pulse repetition frequency, PRF, is illustrated by the data of Figure 8.4, which regard all three types of microbubbles. The results were obtained using a 2.2 MHz transducer emitting pulses of 8 cycles at a temperature of $37°$C. Over the frequency range considered in this test, there is no significant effect of PRF on P_{thr}. The only noticeable and systematic variation appears below 50 Hz, where the time between consecutive arrivals is long enough for Brownian and/or convective motion of the UCAs to change the configuration of the suspension. This change, in turn, produces variations in consecutive time-domain signals which tend to lower the cross-

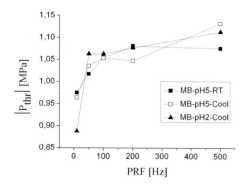

Fig. 8.4. Dependence of P_{thr} on pulse repetition frequency, PRF. These data were obtained at $37°$C, for $N = 8$, and $f = 2.2$ MHz

correlation of the signals. This effect should be considered as an artifact since it amounts to an improper use of the method devised to assess P_{thr}.

Figure 8.5 reports a comparison between values of P_{thr} carried out on suspensions of microbubbles in water and in blood at $37°$C. A correct interpretation of these data must account for the larger attenuation experienced by ultrasonic waves when they propagate in blood rather than in water. In whole blood, the attenuation coefficient is 0.2 dB/(cm · MHz) (**UltraSound – Technology Information Portal** at http://www.us-tip.com), while in water it is roughly two orders of magnitude lower. At 2.2 MHz, the intensity of the insonifying beam is attenuated by 0.44 dB every centimeter traveled in whole blood, which corresponds to an amplitude decrease of 10 percent. Therefore, to generate peak negative pressure fields of identical value in the focal region, the same transmitting transducer must be driven by a higher voltage signal when it operates in blood. All this taken into account, the result of Figure 8.6 concerning the MB-pH5-RT microbubbles suggests that blood has a slightly softening effect on these shells. On the other hand, the data referring to the

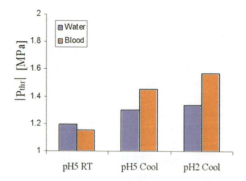

Fig. 8.5. Comparison between the absolute value of P_{thr} in water or in blood. The data were acquired at $37°$C, for $N = 8$, and $f = 2.2$ MHz

microbubbles MB-pH5-Cool and MB-pH2-Cool show a systematic stiffening effect of their shells when interacting with blood. Although no explanation is available at the present time for this observation, its relevance for future applications of these UCAs as drug-carriers cannot be underestimated.

8.4 Conclusions

In this chapter the acoustic and fracture properties of three sets of bubbles that have potential to be used as third generation UCAs have been described. To be usable the bubbles must give an echo and it must be possible to break them within safe pressure limits. The three sets of bubbles described in this book do all provide a signal more than 20 dB above the noise. Compared to other sets of bubbles they are 10 times more effective per particle than those examined by Lathia et al. [5], 100 times than those used by Lavisse et al. [6], but just a third of those used by Wheatley et al. [8]. When comparing the three sets of bubbles, pH2-Cool had a wave attenuation that was about half that of pH5-Cool and pH5-RT.

In clinical practice the temperature of the surrounding media and of the bubbles will be approximately 37°C. In vitro, it is however impractical to carry out all tests at 37°C as it requires heating and control of the surrounding environment. As described earlier, the temperature does not affect the spectra of the backscattered signal, but increases the amplitude of the backscattered signal. The bubbles are also more easily fractured at higher temperatures. Both these changes are beneficial for the clinical use of the bubbles.

To enable measurements of the backscattered signal (for instance in order to produce the image) the ultrasound beam is sent out in small bursts. Each burst contains a few cycles of the ultrasound carrier frequency. When sonified the bubbles change in size with the frequency of the ultrasound. At low pressures the bubbles grow in size and at high pressures they decrease in size. It is at the negative pressure that the bubbles crack and release their content. This is called the peak negative pressure, P_{thr}. If the number of cycles in every burst is increased, a lower pressure is needed to crack the bubbles. This decrease in fracture strength with increasing number of cycles per burst implies that the shell is affected by fatigue. How quickly the bursts are repeated, the pulse repetition frequency, does not affect P_{thr}. The peak negative threshold is of great importance for clinical imaging. Below P_{thr} the UCAs can be imaged non-destructively and above it the UCA's therapeutic payload may be locally delivered by fracturing their shell. The P_{thr} should be high enough so that the bubbles do not break too early, but not so high that the pressure needed to break the bubbles is harmful to the imaged tissue. For safety reasons the mechanical index should be below 1.9 (the mechanical index is defined as the ratio between the peak negative pressure and the square root of the frequency of the ultrasonic wave of interest). When compared to bubbles with thinner shells, the bubbles described in this book fracture at greater MI values but still sufficiently low that no inertial cavitation event in the hosting environment can be observed. Thus, the bubbles tested were all possible to break within the safety limits.

Tests of ultrasound bubbles are for practical reasons most frequently made in degassed water. The propagation of ultrasound in blood is however different from that of water – the attenuation coefficient is about 100 times greater in blood. This will affect the attenuation of the backscattered signal. At 2.2 MHz, the intensity of the insonifying beam is attenuated by 0.44 dB every centimetre travelled in whole blood, which corresponds to an amplitude decrease of 10 percent. Blood does also have a slightly softening effect on the shells of bubbles synthesized at low temperature, while those made at room temperature, which is also the most echogeneic and with the lowest fracture threshold, appear to be insensitive to the nature of the hosting medium. In other chapters the chemistry of PVA polymers and the interaction with blood is described, which in part may explain this difference.

References

1. Cavalieri F, El Hamassi A, Chiessi E and Paradossi G (2005) Stable Polymeric Microballoons as Multifunctional Device for Biomedical Uses: Synthesis and Characterization. Langmuir 21(19):8758–8764
2. Chen WS, Matula TJ, Brayman AA and Crum LA (2003) A comparison of the fragmentation thresholds and inertial cavitation doses of different ultrasound contrast agents. J Acoust Soc Am 113:643–651
3. Church C (1995) The effect of an elastic solid surface layer on the radial pulsations of gas bubbles. J Acoust Soc Am 97(3):1510–1521
4. Grishenkov D, Pecorari C, Brismar TB and Paradossi G (2009) Characterization of acoustic properties of PVA-shelled ultrasound contrast agents: linear properties (Part I). Ultrasound Med Biol. In print
5. Lathia JD, Leodore L and Wheatley MA (2004) Polymeric contrast agent with targeting potential. Ultrasonics 42:763–768
6. Lavisse S, Rouffiac V, Peronneau P, Paci A, Chaix C, Reb P, Roche A and Lassau N (2008) Acoustic characterization of a new trisacryl contact agent. Part I: In vitro study. Ultrasonics 48:16–25
7. Pecorari C and Grishnkov D (2007) Characterization of ultrasound-induced fracture of polymer-shelled ultrasonic contrast agents by correlation analysis. J Acoust Soc Am 122:2425–2430
8. Wheatley MA, Forsberg F, Oum K, Ro R and El-Sherif D (2006) Comparison of in vitro and in vivo acoustic response of a novel 50:50 PLGA contrast agents. Ultrasonics 44:360–367

Chapter 9

Novel Characterization Techniques of Microballoons

Paulo Fernandes, Melanie Pretzl, Andreas Fery, George Tzvetkov, and Rainer Fink

Abstract. With the development of innovative, more complex and ever smaller micro-systems there is a growing need to develop novel characterization techniques that can provide more detailed information about their properties. Fundamental aspects of these techniques are high resolution, measurements in liquid environment, force detection and chemical sensitivity on individual microballoons. In this chapter three characterization techniques based on AFM, RICM and STXM are presented and illustrated with some basic results obtained on PVA-based microballoons.

9.1 Characterization of MBs by AFM

Mechanical properties of microbubbles (MB) are obviously important as they determine stability. As targeted drug delivery agents triggered by ultrasound the microbubbles have to posses enough robustness to avoid membrane rupture and enough elasticity to favor targeted adhesion while still susceptible to ultrasound induced membrane rupture. There is therefore a complicated interplay of different interaction processes that closely depend on the mechanical properties of the MBs and that need to be understood in order to better control their behavior in applications.

Several techniques exist to study mechanical properties of microbubbles [17]. The atomic force microscope (AFM) [4] offers several advantages as it can apply a wide range of forces (from tens of pico to microNewtons) and detect deformations smaller than 1 nm on individual microbubbles in solution and at different temperatures. This renders the technique very attractive to study the mechanical properties of microbubbles. In addition this technique has been successfully used to study mechanical properties of polyelectrolyte multilayer capsules (PEM, [10, 11, 13, 16, 19, 30]) and vesicles [8, 9].

In this section the mechanical experiments performed on MBs (prepared at room temperature and pH 5) will be presented. In a first part the AFM force spectroscopy technique is introduced followed by the most important results obtained.

Paradossi, G., Pellegretti, P., Trucco, A. (Eds.)
Ultrasound contrast agents. Targeting and processing methods for theranostics
© Springer-Verlag Italia, 2010

9.1.1 Experimental setup

An AFM mounted on an inverted optical microscope was used to probe the mechanical properties of individual MBs (Fig. 9.1). During force spectroscopy AFM experiments an individual MB is compressed by a cantilever (moving a piezo) while the deflection of the cantilever is measured by an optical lever (detection of a laser reflected on the tip of the cantilever). With a calibrated cantilever the deflection versus piezo displacement data can be transformed into a force versus deformation curve which gives the mechanical response of the particular MB to the applied force. Force spectroscopy measurements were carried out under water using commercial AFM setups: a Nanowizard (JPK Instruments, Germany) used for the experiments at body temperature and an MFP1D (Asylum Research) for most of the experiments at room temperature. In both setups the AFMs were placed on top of inverted optical microscopes (Zeiss Axiovert 200 for the first and Olympus IX71 for the second) to control the alignment and monitor the MBs during the compression. A colloidal probe [6, 12] (glass bead, diameter ∼50 um, PolyScience inc.) was glued to a tipless cantilever (MikroMash, Spain) with two component epoxy glue (UHU Plus endfest 300, UHU GmbH & Co. KG, Germany) using a micromanipulator (Suttner Instrument Co.). The spring constants of the cantilevers were determined using the thermal noise method [21] or the Sader method [39] (for stiffer cantilevers). To perform a force spectroscopy measurement the MBs have to be reasonably attached to the substrate to avoid slippage during compression. To promote this adhesion a droplet of MBs solution was applied to a PEI coated thin glass slide. The experiments revealed that slippage can still occur if the MBs and probe are not aligned (this should be carefully monitored by use of the inverted optical microscope). Typical compression/retraction cycles were done in 1 second.

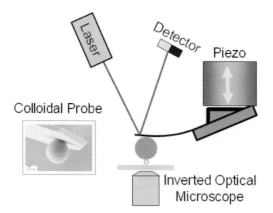

Fig. 9.1. Schematic diagram of the AFM setup used with an Inverted Optical Microscope. A colloidal probe (silica bead) was glued to a tip-less cantilever and the measurements were taken under water

9.1.2 Bubble bursting

In Figure 9.2 the force-deformation curve obtained for a typical MB is presented (the force-deformation curves should be read from right to left). The compression curve (trace, blue) presents a first zone where the force increases with the deformation of the sample until a point where there is a peak followed by a decrease in the measured force. Next the effect of the substrate is observed by an almost vertical increase of the force. The red retraction curve shows that after such deformation the MB does not behave elastically, as expected. We interpret the force peak as the bubble bursting. This scenario is compatible with the optical microscopy images observed (Fig. 9.3) before and after the bursting event: before bursting the air-filled MB has a larger index of refraction gradients (between air and polymer) so it scatters more light (Fig. 9.3, right) than after the bursting where water replaces air and, consequently the water-filled MB presents a smaller index of refraction gradients (only between water and polymer) so it scatters less light (Fig. 9.3, left).

Several MBs were analyzed and presented similar bursting profiles, from which a burst force (peak force in the force-deformation curve) could be measured. In Figure 9.4 a graph of the burst as a function of bubble radius is presented. There is some dispersion, as expected, since we are analyzing indi-

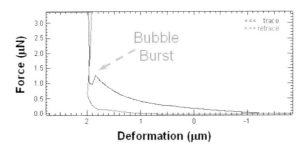

Fig. 9.2. AFM force-deformation curve of a MB (R~2.36 μm), where bubble bursting can be observed

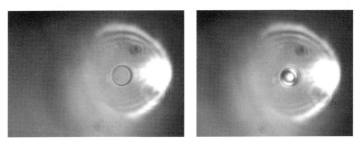

Fig. 9.3. Inverted optical microscope images of a MB, after (left) and before (right) bursting

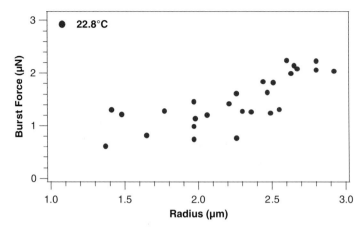

Fig. 9.4. Burst force versus bubble radius at room temperature

vidual MBs that can present an important polydispersity in wall thickness and composition, air pressure. Nevertheless the data seems to indicate an increase of the burst force with bubble radius. One more point regarding the bursting phenomenon: experiments performed with slower compression/retraction cycles lower burst forces were observed. This is also compatible with the bursting scenario due possibly to crack propagation kinetics.

9.1.3 Bubbles after bursting

An important question regarding the bubble bursting is what happens to the bubble afterwards? To elucidate this point force spectroscopy measurements (compression/retraction cycles) were made after the MB bursting was observed. In Figure 9.5 the force versus piezo displacement curves of a MB during (left) and after (right) bursting are presented. In the force curve after bursting we can notice that the force response of the MB starts at approximately the same piezo displacement as in the bursting curve. This means that the MB does not fragment into pieces after burst but actually recovers almost completely its previous shape. We note further that the behavior after burst is reproducible for at least 10 compression/retraction cycles.

9.1.4 Bubbles' stiffness cycle dependence

AFM force spectroscopy can provide much more information about the sample than just the burst force. In particular, the linear slope of the force curve at small deformations can be used as a measure of the stiffness of the MBs wall. By performing several push-pull cycles on the same MB we observed that its stiffness changes. Two different typical behaviors were identified:

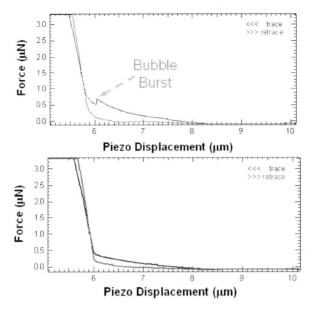

Fig. 9.5. AFM force-deformation curve of a MB (R~2.26 um), during (left) and after (right) bursting

- for soft cycles (with low applied forces, smaller than 50 nN, corresponding to small bubble deformations) the bubble wall stiffness increased (Fig. 9.6, left);
- for hard cycles (with high applied forces, higher than 1μN, corresponding to large bubble deformations) the bubble wall stiffness decreased (Fig. 9.6, right).

We note that in both cases the total cycle time was identical (one second). Several factors can play a role in this phenomenon like water displacement kinetics, visco-elastic effects, introduction of structural defects. Further exper-

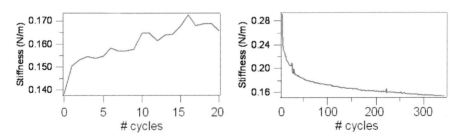

Fig. 9.6. Typical stiffness response of MBs versus number of push-pull cycles under low applied forces (small deformations, left) and high applied forces (large deformations, right)

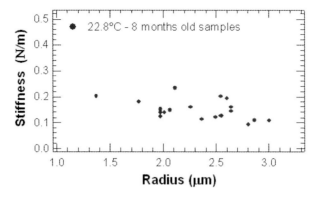

Fig. 9.7. Stiffness versus MB radius at room temperature

iments will be performed to understand this interesting and reproducible behavior.

9.1.5 Bubble stiffness

Since the stiffness of the MBs changes with the number of push-pull cycles it has been subjected to, in order to obtain the stiffness dependence as a function of bubble radius (Fig. 9.7) only the first push-pull cycle was considered. A relatively linear dependence is observed, the dispersion is quite reasonable taking into account the polydispersity of the samples (wall thickness, wall composition) and the fact that the stiffness depends also on the past mechanical history of the bubbles. To determine the Young's modulus of the MB wall from these data an appropriate model would have to be developed, taking into account the wall thickness and composition gradient. Nevertheless, preliminary estimation can be given by using the Reissner model [37, 38] (applicable for thin shells [26], already successfully used to determine the Young's modulus of polymeric microcapsules [10,11,13,16,19,30]). Therefore, by considering an effective wall thickness of 600 nm the effective Young's modulus of the wall material according to the Reissner model is approximately 400 kPa.

9.1.6 Bubble temperature dependence

Taking into account the projected application of the MBs it is important to study their mechanical property dependence with temperature. To this effect several compression/retraction cycles were performed on the same air-filled MB at room temperature and after at body temperature. To avoid bursting, only small deformations were imposed on the MB by using a softer cantilever. Several force-deformation curves were collected and from each one the stiffness (or slope) of the curve in the small deformation (linear) regime were determined. The results at room and body temperature are presented in the

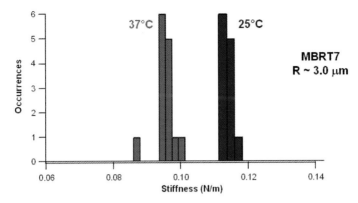

Fig. 9.8. Histograms of MB stiffness at room and body temperature

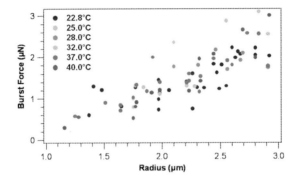

Fig. 9.9. Burst force versus MB radius at different temperatures

histogram of Figure 9.8 and clearly show that the MBs became softer at body temperature. We remark that, as seen before, a sequence of soft push-pull cycles (like the ones carried out during this experiment) should increase the stiffness of the MB. Since we observed a stiffness decrease from room to body temperature it must be due to the temperature change and not to the push cycle sequence. The temperature softening effect (of around 10 %) might be underestimated as there is probably some stiffening due to the consecutive push cycles.

In Figure 9.9 we present the burst force versus MB radius at different temperatures. The dispersion is again expected, but it is interesting to note that the bursting force seems to be rather independent from temperature. Since a temperature induced softening of the wall material has been put into evidence (and the softness of the wall material should influence its bursting point) there might be an interplay of two effects that cancel each other out.

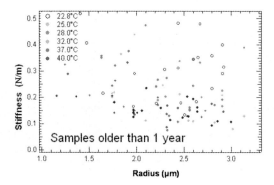

Fig. 9.10. Stiffness versus MB radius at different temperatures for MB samples older than 1 year

9.1.7 Sample age

The stiffness of the MBs was also measured at different temperatures for samples prepared more than one year before the experiment (Fig. 9.10). Instead of linear behavior as in Figure 9.7 a more or less random distribution is observed. This result suggests that the polydispersity is significantly increased with sample age (this is compatible also with the fact that the stiffness of the bubbles depends on its previous history: the older the samples the more history it has and the less predictable the stiffness values are).

9.1.8 Adhesion forces

In addition to the mechanical properties of the microballoons it is of course essential for their application, in particular as drug delivery systems, to study the interactions between microballoons and surfaces. An important task is then to control the interactions between the MB and the surrounding tissue. An excellent tool to study adhesion properties between microballoons and substrates is the colloidal-probe AFM combined with an optical microscope [13, 36], that was already introduced for the measurement of mechanical properties (Fig. 9.1).

The optical microscope is, for these measurements, used in the reflection interference contrast microscopy mode (RICM). This technique developed by Sackmann and Co-workers is very useful for investigating adhesion areas of particles based on the interference of reflected light [35]. Microbubbles that are adhered to a glass substrate show in the RICM mode typical bright spots surrounded by an interference pattern. This pattern originates from reflected light from the microballoon shell and the substrate surface that is interfering constructively or deconstructively. In Figure 9.11 a typical interference pattern of adhered MBs with a size of 4 μm is displayed. In this experiment the polymer shell of the microballoons is modified with hyaluronic acid, thus the

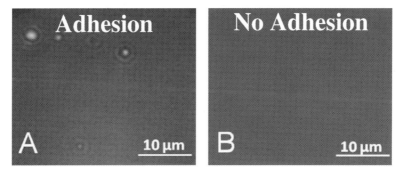

Fig. 9.11. (a) Typical RICM result of adhered MB: Bright spots and Newton fringes. (b) MB exposed to a repulsive substrate: No adhesion in the RICM visible

zeta potential increases to -20 mV ± 1.5 mV. The negative charged MBs stick to positive charged substrates and adhesion is observed. If the same microballoons are exposed to negatively charged substrates (Fig. 9.11b) no adhesion and no bright spots are visible with RICM.

To quantify adhesion forces with RICM in detail, an apparent contact area between MB and substrate can be calculated. The contact or adhesion area for MB is considered as the constant grey region in the RICM images. For the determination of the area an intensity profile is extracted and the diameter of the contact region can be determined. The potential of the RICM technique is its combination with colloidal probe AFM. External forces can be applied onto the MB in a liquid environment and the adhesion forces measured with AFM can be correlated with the change of the contact area followed by RICM.

When the colloidal probe detaches from the MB a pull-off force (adhesion force) can be determined, shown in Figure 9.13a. For increasing external forces

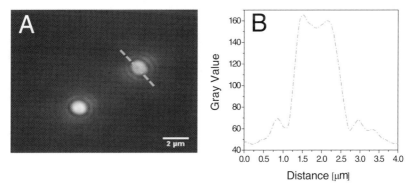

Fig. 9.12. (a) The adhesion area of the microballoons is the bright spot that is surrounded by an interference pattern. (b) Extracted intensity profile of the interference pattern

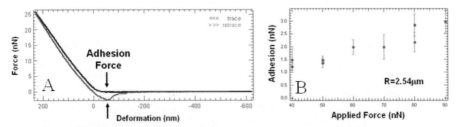

Fig. 9.13. (a) AFM force-deformation curve, adhesion between Probe and MB can be measured. (b) Adhesion between MB and Probe versus applied force

an increase in the observed pull-off forces is displayed in Figure 9.13b, due to the enlarged adhesion area.

With this technique several individual microballoons can be tested. The observed adhesion forces are influenced by the size of the particles, the shell thickness and the air/water content of the MB. These results show that the MBs adhesion forces can be measured with the used AFM setup and that it is a very interesting and versatile technique to probe this system.

9.1.9 Adhesion arrays

The directed particle adhesion onto patterned substrates is interesting for a wide range of applications in the field of combinatorial chemistry, the buildup of sensor arrays or optical materials. The self assembly of MB on structured substrates is of great interest, because it is an inexpensive approach to set up controlled arrangements of MB over a large area that can be used for serial testing of microballoons or used as test substrates for cell exposure experiments. The observation and quantification of microballoon adhesion to specific substrates allowed us to progress towards this goal. For the preparation of patterned substrates in addition to lithographic methods there are also various "soft lithographic" techniques available: microcontact printing [5, 35] polymer on polymer stamping (POPS) [28], wrinkling [18, 29]. Several studies have been devoted to the selective deposition of polyelectrolytes cells [23, 33], proteins [20, 22] and microcapsules [27, 31, 40].

To direct the adhesion of negatively charged MB (modified with hylaluronic acid) a patterned substrate with different charge densities can be used. In Figure 9.15a a line pattern was transferred via microcontact printing (Fig. 9.14) onto a multilayer coated glass substrate. This process is relatively easy to use, reproducible and allows the production of a very well defined broad range of patterns.

The bright lines in Figure 9.15a are positively charged, while the dark background exhibits a negative charge. Negatively charged MB adhere preferential to the line pattern (Fig. 9.15b). Due to the used polyelectrolyte a strong or weak adhesion of the MB onto the pattern can be controlled.

9.1.10 Conclusions

The AFM setup used allows not only the observation (through force curves corroborated with optical microscopy images) of individual MB bursting but also and quite interestingly the quantification of the burst force. Both wall stiffness and burst force seem to depend linearly on the MB radius. Measurements at both room and body temperature revealed that the MBs became softer at body temperature but the burst force is relatively temperature independent. A significant increase in wall stiffness polydispersity with sample age was put into evidence.

The presented setup, colloidal probe AFM and an optical microscope in RICM mode, offers a versatile technique to study adhesion forces on different substrates and under external forces. Adhesion forces increase with increasing external applied forces and the dependence on the deformation of the soft particles and the resulting change of the contact area can be studied with the setup in detail. The adhesion forces are influenced by various parameters like the size of the MB, the shell thickness and the MB air/water content. The quantification of adhesion energies between MB and different substrates was crucial for the setup of MB arrays. These arrays are interesting substrates for the serial testing of microballoons and cell exposure experiments.

9.2 Characterization of MBs with STXM

Zone-plate based scanning transmission soft X-ray microspectroscopy (STXM) is a rapidly developing analysis technique which makes use of the advantages of high brilliance synchrotron radiation [1, 2, 24, 32]. In STXM synchrotron, X-ray radiation is focused by a Fresnel zone plate and the sample is raster-scanned through the focal point while recording the intensity of transmitted X-rays. Thus, a 2D image is formed like in other scanning probe techniques.

A schematic diagram of the STXM setup is shown in Figure 9.16. The Fresnel zone plate (FZP) serves as a demagnifying/focusing diffractive element. FZP is a circular diffraction grating of alternate transparent and opaque zones. Higher diffraction order beams are blocked by a pinhole, which serves

1.   2.   3.   4.

Fig. 9.14. Schematic diagram of the micro-contact printing process: (1) Incubation of the elastomeric stamp with an aqueous solution of a fluorescent-labeled polyelectrolyte. (2) After rinsing and drying a thin monolayer is obtained on the stamp surface. (3) Transfer of the stamp onto a polyelectrolyte multilayer coated substrate. (4) After removal of the stamp a patterned substrate is obtained

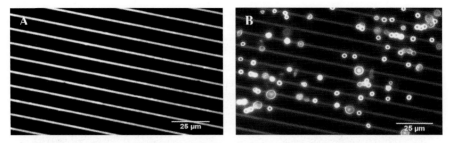

Fig. 9.15. (a) Line pattern transferred with microcontact printing to a polyelectrolyte multilayer. (b) Selective adhesion of microballoons on patterned substrate

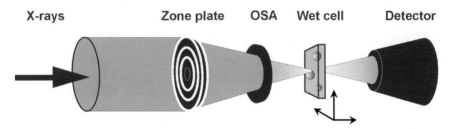

Fig. 9.16. Scheme of a scanning transmission X-ray microscope (STXM) showing Fresnel zone plate (FZP) producing a diffraction limited focus, order selecting aperture (OSA) selecting only the first order focus, the wet cell, and the detector measuring the transmitted intensity

as an order-sorting aperture (OSA). Based on the near-edge X-ray absorption fine structure (NEXAFS) contrast, STXM can be used for elemental and chemical imaging to determine the molecular composition in the sample. Thus, by measuring the energy dependent transmission in the focus of the X-ray beam, the experiment provides chemical and sub-40 nm structural data which can be directly correlated. Furthermore, by operating in the "water window" spectral region between the carbon and oxygen K-edge absorption edges (about 285 eV to 535 eV), one can study samples in up to 10 mm of water or ice. STXM provides higher spatial resolution than the confocal laser scanning microscope (CLSM), at present a widely used technique for the investigation of microcapsule systems, which makes the X-ray microscopy very advantageous for obtaining new insights into the nanoscale assemblage of such materials. Besides, additional sample preparation like, for example, fluorescence labeling, is not used in STXM, since it utilizes the spectroscopic contrast which allows for quantitative chemical analysis.

For STXM measurements we used the so-called "wet cells" where approximately 1 μL of well homogenized microballoon (MB) water suspension was sandwiched between two 100 nm thick Si_3N_4 membranes (Silson Ltd, UK), which were then sealed with silicone high-vacuum grease to maintain the water environment during the experiment. The MBs were imaged in transmission

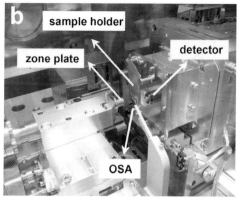

Fig. 9.17. Photographs of the PolLux-STXM at beamline X07DA at the Swiss Light Source: (**a**) microscope chamber in the experimental hutch and (**b**) main elements of the setup [3]

mode in a helium atmosphere using the PolLux-STXM microscope at the Swiss Light Source (SLS), Paul Scherrer Institut (see Fig. 9.17). The transmitted photon flux was measured using a photomultiplier tube (Hamamatsu 647P). The SLS storage ring runs at 2.4 GeV and "top-up" operation mode which guarantees a constant electron beam current of 400 ± 1.5 mA. The PolLux-STXM uses linearly polarized X-rays from a bending magnet in the photon energy range between 200 eV and 1200 eV and it provides a spatial resolution better than 40 nm and spectral resolving power $E/\Delta E > 5000$ (at the N K-edge, approximately 400 eV) [3]. Images were recorded at selected energies through the O 1 s region (510–560 eV). Oxygen K-edge NEXAFS spectra were acquired in the so-called line-scan mode, i.e. the transmitted intensity signal was recorded while a line trajectory was scanned across a part of the sample for each photon energy through the spectrum. The line scans were performed with 0.1 eV energy steps from 520 eV to 560 eV and the NEXAFS spectra were normalized to unity at 560 eV.

So far, we have concentrated on two aspects: first, STXM was used to monitor the interior of the microballoons, since the absorption contrast close to the O K-edge is superior to distinguish between air and water-filled MBs. The second aspect is concerned with the stabilizing shell of the MBs: from the 2D projection of spherical MBs the thickness profile of the shell can be quantitatively estimated.

Two STXM transmission images of three PVA-based microballoons in the water environment recorded at 520 eV and 550 eV (i.e. below and above the 1 s absorption threshold of oxygen) are presented in Figure 9.18. STXM microscopy is based on the contrast given by the absorption coefficients of the species; the transmitted photon intensity, through the material, depends on the thickness, density and atomic number of each component according to Lambert-Beer's law. Taking into account the calculated transmission curves in

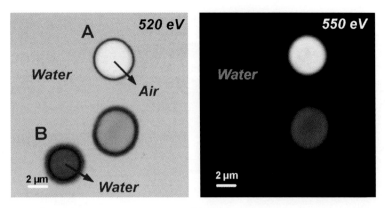

Fig. 9.18. STXM transmission images of MBs in a water environment recorded at $h\nu = 520$ eV and $h\nu = 550$ eV (scanned image size: 20×20 μm^2)

the oxygen K-edge region of water, PVA and air components [34], it becomes clear that the STXM image at $h\nu = 520$ eV shows the PVA shells of the MBs while the water and air absorption is weak compared to the carbonaceous material. Above the oxygen K-edge ($h\nu = 550$ eV), a water environment and PVA strongly absorb the X-rays while air shows approximately one order of magnitude higher transmission of the X-rays. Consequently the gas-filled cores of the particles appear brighter. Additionally, the core of the MB B shows essentially no contrast compared to the water background. This unambiguously suggests that air was released through the membrane and the particle is water-filled. Thus, the contrast variations in the STXM transmission images below and above the O K-edge provide direct evidence of the MB gas interior.

In order to gain insight into the chemical composition of the interior of the MBs, oxygen K-edge NEXAFS spectroscopy was applied. Absorption spectra extracted from line scans across the inner part of particles A and B are compared in Figure 9.19. A NEXAFS spectrum taken from a water reference is also shown in Figure 9.19. A reference spectrum of the "water-free" polymeric shell was also obtained after drying the wet cell in the STXM chamber overnight. The oxygen K-edge NEXAFS spectrum taken from a microballoon-free volume of the sample exhibits the typical absorption features of liquid water. The water spectrum starts at 535.4 eV followed by the characteristic two-peak broad structure in the energy range of 537–543 eV [41]. The presence of a small feature at about 532 eV is most probably due to some organic contamination in the water (see below). The O K-edge spectrum from MB B shows essentially the same resonances as the water spectrum except a small intensity decrease of the feature at 535.4 eV and an intensity increase of the peak at 532.1 eV. This result unambiguously underlines the presence of water inside particle B. In contrast to the spectra of water and particle B, the NEXAFS spectrum of microballoon. A demonstrates a line shape which is very similar to that of a particle in a dry state (see Fig. 9.19). The latter two spectra

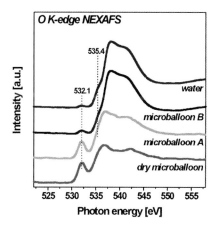

Fig. 9.19. Oxygen K-edge NEXAFS spectra of the surrounding water, the interior parts of MBs A and B (Fig. 9.18), and of another MB analyzed in the completely dehydrated state

show a strong resonance at 532.1 eV which is assigned to the O1 s $\rightarrow \pi*_{(C=O)}$ transition originating from the carbonyl groups of the telechelic PVA shell. Furthermore, the water peak at 535.4 eV is absent in these spectra while the main O1 s $\rightarrow \sigma*$ resonance appears at around 537 eV. Hence, the NEXAFS spectrum of MB A shows only the resonances typical for the telechelic PVA shell. This result strongly suggests that MB A is air-filled. The STXM images presented in Figure 9.18 corroborate this conclusion. The air in the microballoons appears much lighter than the water background in the STXM image at 550 eV while the water-filled particle B is indistinguishable at this energy.

Very recently, a quantitative analysis of the STXM transmittance profiles of the MBs where the X-ray beam resolution and a third order polynomial radial membrane absorption function were taken into account, was reported [43]. In summary, the model is based on the Lambert-Beer expression for the transmitted monochromatic X-rays through a three-component system (encapsulated air, PVA-based shell and the surrounding water in the wet cell) and extends the previous quantitative studies of water-filled polymeric microcapsules using full-field transmission X-ray microscopy (TXM) [15,25]. From the proposed fitting procedure the MBs' physical parameters such as radius, wall thickness and wall absorption can be determined with unprecedented high resolution. This analytical model opens new applications for quantitative characterization of multicomponent microcapsule materials by means of STXM.

STXM transmission image at 282 eV of freeze-dried PVA-based MBs deposited onto a Si_3Ni_4 membrane is depicted in Figure 9.20. At this photon energy (below the C1s absorption edge) the contrast in the image originates from the topographical (thickness) differences in the microcapsules. As one sees, the image clearly shows the deformed polymeric shells of the dry parti-

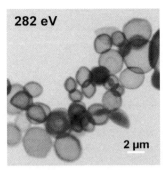

Fig. 9.20. STXM transmission image at hν = 282 eV of the freeze-dried MBs

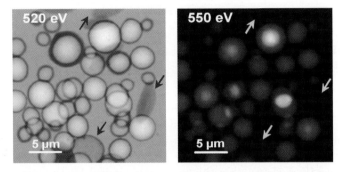

Fig. 9.21. STXM transmission images recorded at hν = 520 eV and hν = 550 eV of freeze-dried MBs suspended in water. The statistical analysis has shown that approximately 80 % of the MBs in the wet cells are air-filled

cles. The MBs were spectromicroscopically characterized after suspending the freeze-dried material in water. STXM images taken at 520 eV and 550 eV of the MBs suspension in a wet cell are shown in Figure 9.21. The changes in the freeze-dried MBs morphology are noticeable as one can see in the image at 520 eV. Furthermore, the contrast variations in the STXM images below and above the O *K*-edge (see Fig. 9.21) explicitly demonstrates the gas interior of the MBs. There are also several broken shells present (indicated with arrows in Fig. 9.21) which are indistinguishable from the water medium in the STXM image at 550 eV.

The present examples demonstrate the potential of the STXM technique for characterization of microballoon and microcapsule systems. STXM imaging below and above the oxygen *K*-edge (520 eV and 550 eV) and NEXAFS spectroscopy can provide unique information on the composition of the MBs in water. Furthermore, with STXM it is possible to gain more detailed information on the variations in chemical structure of MBs subjected to external stimuli like temperature, mechanical forces or light irradiation. Thus, the analytical X-ray microscopy might be vital for the development of modern biochemical devices and applications in drug delivery and as ultrasound contrast agents. These first results will be extended to different types of hollow and solvent-filled therapeutic microcapsules.

References

1. Ade H, Zhang X, Cameron S, Costello C, Kirz J and Williams S (1992) Science 258:972
2. Ade H and Hsiao B (1993) X-ray Linear Dichroism Microscopy. Science 262:1427
3. Ade H and Hitchcock AP (2008) NEXAFS microscopy and resonant scattering: Composition and orientation probed in real and reciprocal space. Polymer 49:643
4. Binnig G, Quate CF and Gerber C (1986) Atomic Force Microscopy. Phys Rev Lett 56:930
5. Booth BL (1993) Polymers for electronic and photonic applications. In: Wong CP (ed). Academic Press, Boston, pp. 549–597
6. Butt HJ (1991) Measuring Electrostatic, Van der Waals, and Hydration Forces in Electrolyte-Solutions with an Atomic Force Microscope. Biophys J 60:1428
7. Déjugnat C, Köhler K, Dubois M, Sukhorukov GB, Möhwald H, Zemb T and Guttmann P (2007) Membrane densification of heated polyelectrolyte multilayer capsules characterized by soft X-ray microscopy. Adv Mater 19:1331
8. Delorme N and Fery A (2006) Direct method to study membrane rigidity of small vesicles based on atomic force microscope force spectroscopy. Phys Rev E 74:030901
9. Delorme N, Dubois M, Garnier S, Laschewsky A, Weinkamer R, Zemb T and Fery A (2006) Surface immobilization and mechanical properties of catanionic hollow faceted po
10. Dubreuil F, Elsner N and Fery A (2003) Elastic properties of polyelectrolyte capsules studied by atomic force microscopy and RICM. Europhys J E 12:215–221
11. Dubreuil F, Shchukin DG, Sukhorukov GB and Fery A (2004) Increasing Stiffness and Toughness of Polyelectrolyte-capsules by YF3-nanoparticle modifying: an AFM study. Macromol rapid Comm 25:1078
12. Ducker WA, Senden TJ and Pashley RM (1991) Direct Measurement of Colloidal Forces Using an Atomic Force Microscope. Nature 353:239
13. Elsner N, Dubreuil F and Fery A (2004) Tuning of microcapsule adhesion by varying the capsule-wall thickness. Phys Rev E 69:031802
14. Elsner N, Dubreuil F, Weinkamer R, Fischer F.D., Wasicek F and Fery A (2006) Mechanical properties of freestanding polyelectrolyte capsules. Prog Coll Polym Sci 132:117–132
15. Fernandes P, Tzvetkov G, Fink R, Paradossi G and Fery A. (2008) Langmuir 24(23): 13677
16. Fery A, Dubreuil F and Möhwald H (2004) Mechanics of artificial microcapsules. New Journal of Physics 6:18
17. Fery A and Weinkamer R (2007) Mechanical properties of micro- and nanocapsules: Single-capsule measurements. Polymer 48:7221–7235
18. Hammond PT and Whitesides GM (1995) Formation of Polymer Microstructures by Selective Deposition of Polyion Multilayers Using Patterned Self-Assembled Monolayers as a Template. Macromolecules 28(22):7569–7571

19. Heuvingh J, Zappa M and Fery A (2005) Salt induced softening transition of polyelectrolyte multilayer capsules. Langmuir 21:3165
20. Howell SW, Inerowicz HD, Regnier FE and Reifenberger R (2003) Patterned protein microarrays for bacterial detection. Langmuir 19(2):436–439
21. Hutter JL and Bechhoefer J (1993) Calibration of Atomic-Force Microscope Tips. Rev Sci Instrum 64:1868
22. Kam L and Boxer SG (2001) Cell adhesion to protein-micropatterned-supported lipid bilayer membranes. Journal of Biomedical Materials Research 55(4):487–495
23. Kane RS, Takayama S, Ostuni E, Ingber DE and Whitesides GM (1999) Patterning proteins and cells using soft lithography. Biomaterials 20(23–24):2363–2376
24. Kirz J and Rarback H (1985) Rev Sci Instrum 56:1
25. Kohler K, Déjugnat C, Dubois M, Zemb T, Sukhorukov GB, Guttmann P and Möhwald H (2007) J Phys Chem B 111:8388
26. Koiter WT (1963) A spherical shell under point loads at its poles (ed). Macmillan, New York, pp. 155–170
27. Krol S, Nolte M, Mazza D, Magrassi R, Diaspro A, Gliozzi A and Fery A (2005) Encapsulated Living Cells on Microstructured Surfaces Langmuir 21:705
28. Kumar A and Whitesides GM (1994) Patterned Condensation Figures as Optical Diffraction Gratings. Science 263:60–62
29. Lu C, Möhwald H and Fery A (2007) A lithography-free method for directed colloidal crystal assembly based on wrinkling. Soft Matter 3:1530–1536
30. Müller R, Köhler K, Weinkamer R, Sukhorukov GB and Fery A (2005) Melting of PDADMAC/PSS capsules investigated with AFM Force Spectroscopy. Macromolecules 38:9766–9771 lyhedrons. J Phys Chem B 110:1752
31. Nolte M and Fery A (2004) Coupling of individual polyelectrolyte capsules onto patterned substrates. Langmuir 20(8):2995–2998
32. Nolte M and Fery A (2004) Microstructuring of polyelectrolyte coated surfaces for directing capsule adhesion. IEEE Transactions on Nanobioscience 3(1):22–27
33. Pretzl M, Schweikart A, Hanske C, Chiche A, Zettl U, Horn A, Böker A and Fery A (2008) A lithography-free pathway for chemical microstructuring of macromolecules from aqueous solution based on wrinkling. Langmuir 24:12748–12753
34. Raabe J, Tzvetkov G, Flechsig U, Boge M, Jaggi A, Sarafimov B, Quitmann C, Vernooij MGC, Huthwelker T, Ade H, Kilcoyne D, Tyliszczak T and Fink R (2008) PolLux: A new Beamline for Soft X-Ray Spectromicroscopy at the SLS. Rev Sci Instrum 79:113704
35. Raedler J and Sackmann E (1992) On the measurements of weak repulsive and frictional colloidal forces by reflection interference contrast microscopy. Langmuir 8:848–853
36. Raichur A, Vorös J, Textor M and Fery A (2006) Adhesion of Polyelectrolyte Multilayer Capsules through specific biotin-streptavidine interactions. Biomacromolecules 7:2331–2336
37. Reissner E (1946) Stresses and small displacements of shallow spherical shells. J Mat Phys 25:80
38. Reissner E (1946) Stresses and Small Displacements of Shallow Spherical Shells, 2^n. Journal of Mathematics and Physics 25:279–300
39. Sader JE (1998) Frequency response of cantilever beams immersed in viscous fluids with applications to the atomic force microscope. J Appl Phys 84:64–76

40. Saravia V, Nolte M, Küpcü S, Pum D, Fery A, Sleytr UB and Toca-Herrera J (2007) Bacterial protein patterning by micro-contact printing of PLL-g-PEG. J Biotechnol 130:247–252
41. Tzvetkov G, Graf B, Fernandes P, Fery A, Cavalieri F, Paradossi G and Fink RH (2008) Soft Matter 4:510
42. Xia Y and Whitesides GM (1998) Softlithographie. Angew Chem Int Ed 37:551
43. Zubavichus Y, Yang Y, Zharnikov M, Fuchs O, Schmidt TH, Heske C, Umbach E, Tzvetkov G, Netzer FP and Grunze M (2004) Chem Phys Chem 5:509

Chapter 10

A Comparative Analysis of Contrast Enhanced Ultrasound Imaging Techniques

Andrea Trucco, Marco Crocco, Anastasis Kounoudes, and Claudia Sciallero

Abstract. One of the most important issues in the field of ultrasound medical imaging using contrast agents is the development of techniques able to separate the response of contrast media from the response of biological tissues. In the literature, one can find various solutions concerning the use of multiple transmitted signals and the combination of the related echoes. Moreover, such multi-pulse techniques can be combined with a coded excitation in order to increase both the contrast to tissue ratio and the signal to noise ratio. The effectiveness of these techniques depends on many parameters concerning the medical ultrasound device and the physical scenario involved. In particular, some undesired effects always present in real conditions, like the signal distortion caused by the hardware equipment and by non-linear tissue propagation, the thermal noise, and the tissue movements can significantly reduce the theoretical performance. In this chapter a simulation tool is proposed that allows one to calculate the backscattered echo from a population of contrast agents immersed in a biological tissue, considering all the mentioned effects. With this tool, an assessment of the performance of a set of multi-pulse techniques has been carried out under realistic working conditions, showing that sharp differences occur among the techniques, depending on the specific setup considered.

10.1 Introduction

Over the past decade, ultrasound contrast agents (UCAs), in the form of tiny gas bubbles, have been introduced to improve the quality of ultrasound medical images. To exploit these contrast media, research has been actively focused on developing effective imaging techniques to emphasize the contrast agent echoes and, at the same time, abate the surrounding biological tissue echoes. Such techniques rely on the non-linear properties of the microbubbles: when a microbubble is insonified with an ultrasonic pulse having a moderate pressure (e.g., 80 kPa) and the right frequency, the microbubble begins to oscillate in a non-linear way, backscattering an echo whose spectrum is wider than the one of the impinging pulse due to the presence of harmonic components [16]. On the contrary, at the same pressure, a tissue behaves mainly as

Paradossi, G., Pellegretti, P., Trucco, A. (Eds.)
Ultrasound contrast agents. Targeting and processing methods for theranostics
© Springer-Verlag Italia, 2010

a linear system, backscattering an echo made up of the same spectral components characterizing the transmitted signal.

In the literature, several techniques based on the transmission of a sequence of pulses (multi-pulse techniques) [4,11,12] have been proposed. The sequence is made up of differently scaled versions of the same pulse. In reception, the echoes related to all transmitted signals are opportunely multiplied by specific amplitude coefficients and summed up in a post-processing phase. The transmission and reception coefficients are designed so as to allow complete cancellation of every linear contribution, while the desired non-linear components are preserved. In this way the ratio between the contrast-agent and tissue echoes, i.e., the contrast-to-tissue ratio (CTR) can be considerably enhanced.

Moreover, in recent years, coded excitation has been proposed in clinical ultrasound [3] to increase the signal to noise ratio (SNR, where noise accounts for the thermal noise in the electronic circuits of the ultrasound scanner) and consequently the penetration depth. Coded excitation implies the transmission of long signals, in which the code is embedded, and a matched filter introduced into the reception chain to recover the axial accuracy and remove the code. In this chapter frequency-modulated codes [10], based on the transmission of a linear chirp pulse, (i.e., a pulse with an instantaneous frequency changing linearly over time), are taken into account.

If applied to UCAs, the chirp pulse makes it possible to improve the SNR and to increase the CTR. In fact, the response from bubbles depends on both the peak pressure and the pulse duration. This happens because the effects of the excitation signal accumulate over pulse length [1,2] and, therefore, the chirp produces a larger bubble wall excursion with respect to a conventional continuous wave (CW) pulse with the same bandwidth (obviously, with a much shorter duration) and the same peak pressure. As the bubble wall excursion is directly correlated to the generation of harmonic components, a CTR increase can be obtained if the received signal is duly processed.

The multi-pulse techniques can be combined with the chirp coding [6] in order to achieve an increase in CTR and an improvement in SNR, with respect to those characterizing the multi-pulse techniques using CW excitation. In this case the matched filter devoted to recover the axial resolution has to be tuned according to the non-linear component preserved by the multi-pulse technique adopted.

The performance of all these techniques strongly depends on the specific working conditions, including the ultrasound device characteristics and the physical properties of the UCAs population and of the biological tissue. In particular, the theoretical performance may be sharply reduced by a series of undesired effects [5] like tissue movements during the transmission of the pulse sequence, the non-linearities produced by the transmission chain and by the propagation through the tissue, and the thermal noise introduced mainly in the reception chain.

In order to carry out a comparative analysis of the different multi-pulse techniques in realistic working conditions, a comprehensive simulation tool has been developed by the authors. Such a tool allows to simulate all the transmission/reception ultrasound chains and the interaction of the ultrasound signals with the contrast agent population immersed in biological tissue, including all the previously mentioned undesired effects. In this chapter the structure of the simulation tool is extensively described and the results concerning the performance of different techniques are reported and analyzed.

These activities have been carried out in the context of the European Commission TAMIRUT (TArgeted MIcrobubbles and Remote Ultrasound Transduction) project whose aim is to derive the microbubble volumetric concentration working on the signals remotely acquired by means of an ultrasound scanner. Since the preliminary step is devoted to separation of the signal components due to the UCAs from that due to the tissue, the simulation tool can dramatically help in choosing the most suitable signal processing technique, limiting the need of time-consuming in vitro and in vivo tests. Moreover, as explained in another chapter of this book, the tool plays a fundamental role in building a synthetic set of ultrasound signals which is necessary for the concentration estimation task.

10.2 Signal processing methods

10.2.1 Multi-pulse techniques

The most useful characteristic of multi-pulse techniques is their ability to eliminate the linear contribution from the response of a given scene, and to select and preserve specific non-linear components even if they spectrally overlap to the linear component [12]. The differences between the techniques depend on the non-linear order/orders preserved, which are a function of the number of pulses emitted and of the particular amplitude coefficient and phase assigned to each transmitted and received pulse.

The multi-pulse techniques considered in this chapter, selected to work with different non-linear orders and chosen from the most common and appreciated [4, 11, 12], are the following: Pulse Inversion (PI), two Contrast Pulse Sequences (CPSs) based on the transmission of three and four pulses (referred to as CPS3 and CPS4, respectively). In PI, a couple of identical pulses, opposite in sign, are transmitted and the corresponding echoes are summed. It can be easily demonstrated [12] that only the even orders are retained. Among them, only the second order typically falls inside the transducer bandwidth. In CPS3, a given pulse waveform is transmitted three times scaled in amplitude by the coefficients 0.5, −1, and 0.5, respectively. The corresponding echoes are summed. In this case, one can verify that both the second- and third-order terms are preserved, while the linear one is cancelled. It is worth noting that the third-order non-linearity gives rise to two spectral components: one centered at three times the fundamental frequency and the other centered at the fundamental frequency. In CPS4, the amplitude coefficients 0.5, −1, 1, and -0.5 are used to scale the pulse waveform and to transmit it four times. The corresponding echoes are weighted by the coefficients -2, -1, 1, and 2 and summed. In this case, both the linear and second-order terms are deleted, whereas the third-order term is preserved.

10.2.2 Chirp coding

Chirp excitation implies the transmission of a long frequency-modulated burst into a biological tissue. In the signal processing theory applied to radar and sonar [8, 13], the chirp pulse is well known as it makes it possible to obtain a processing gain (i.e., an increase in SNR, when the received pulse is added to white independent Gaussian noise) equal to its TB product, B being the bandwidth and T the duration time. While a conventional CW pulse has a TB product equal to 1, a chirp can have a TB product significantly higher than 1. However, to achieve the processing gain, the received chirp should be processed by a matched filter. This operation, referred to as pulse compression, increases the SNR and reduces the chirp duration time, producing an output pulse with a time length of about $1/B$ [8, 13], independent of the initial duration T.

The transmission of a chirp pulse not only makes it possible to improve the SNR, surely an important feature in several applications to combat thermal noise, but also shows the following crucial characteristic. The response from the bubbles depends on the peak pressure amplitude and the pulse duration. This happens because, if one considers the bubbles dynamic like a dampened mass-spring system, the effects of the excitation signal accumulate over the pulse length [2]. The chirp excitation produces a larger bubble wall excursion with respect to an equivalent (in terms of bandwidth and peak pressure) CW pulse and, consequently, an echo that presents a higher non-linear response. In fact, the bubble wall excursion is directly correlated to the generation of harmonic components in the backscattered echo [15]. This is a very important advantage that, in turn, can yield an increase in CTR, also considering that the level of tissue harmonics due to non-linear propagation depends only on peak pressure, not on pulse energy [7].

10.2.3 Combining chirp coding with multi-pulse techniques

The advantages brought by chirp excitation applied to UCAs, in terms of CTR and SNR increase, can be fully exploited by combining it with a multi-pulse technique. The general scheme consists of transmitting a series of chirp pulses weighted by the proper coefficients, according to the multi-pulse technique chosen. After the sum of the related echoes in reception, a matched filter is applied to the resulting signal in order to recover the axial resolution. Since the linear component is cancelled out by the sum of the echoes, the matched filter acts on one of the non-linear components retained. Such a general scheme has been applied to both CPS3 and PI.

The combination of the chirp pulse with CPS3 (Chirp CPS3) has been proposed by the authors in [6]: in this case the matched filter acts on the spectral component centered around the fundamental, related to the third order non-linearity.

Differently, the combination of the chirp coding with PI (Chirp PI) relies on the spectral component centered around the second harmonic, related to the second order non-linearity. Therefore, the matched filter has to be properly tuned around the right frequency, as done in [2]. In particular, the filter should have an impulse response with an instantaneous frequency that is twice that of the traditionally matched filter, at every time instant.

10.3 Simulation

In this section, a detailed description of the simulation tool developed by the authors is provided. This tool allows to simulate the array transmission of a wideband single pulse or a sequence of pulses, the propagation along the body tissues (not perfectly linear), the interaction with the bubble population and the tissues contained inside a given region, the array reception of the related echoes, the focused beamforming, and the post-processing phase. Effects such as thermal noise, harmonic distortion (mainly related to the hardware characteristics), and body motion have been included and can be easily tuned. Therefore, the simulation tool enables assess to the functioning of a given signal processing scheme under different operative conditions, e.g., varying the bubble concentration, the bubble position, the array transmission and reception options, and the magnitude of the undesired effects.

10.3.1 Bubble response

Several dynamic models have been developed in recent decades that are able to predict the backscattered pressure from a single microbubble in response to an arbitrary ultrasound pulse excitation. Among them, the Marmottant model [9] has been adopted: such a model is very complete and accurate, allowing one to take into account three different situations for the bubble shell: the elastic state, the ruptured state, and the buckled state [9].

The simulation of the whole echo from a bubble population has been carried out through the coherent sum of the echoes related to each bubble. If the bubble concentration is low, as it is normally expected in clinical conditions, the second order effects due to the interaction among different bubbles can be considered negligible [14]. Therefore, it is assumed that the pressure experienced by each bubble is due only to the transmitted pulse and not to the echoes of the nearest bubbles. Moreover, the low bubble concentration induces a negligible additional absorption of the transmitted pulse, thus no shadowing effect has been assumed.

10.3.2 Tissue propagation

The propagation of the transmitted pulse inside the human body has been modelled through the Khokhlov-Zabolotskaya-Kuznetsov (KZK) equation [7]: it is a partially derivative differential equation that takes into account diffraction of the sound beam, tissue absorption due to thermoviscous attenuation and non-linear propagation of the acoustic wave. The equation for the pressure as a function of time and space can be written as follows:

$$\frac{\partial^2 p}{\partial z \partial t} = \frac{c_0}{2}\left(\frac{\partial^2 p}{\partial x^2} + \frac{\partial^2 p}{\partial y^2}\right) + \frac{\delta}{2c_0^3}\frac{\partial^3 p}{\partial t^3} + \frac{\beta}{2\rho_0 c_0^3}\frac{\partial^2 p}{\partial t^2}, \qquad (10.1)$$

where p is the acoustic pressure; z is the spatial coordinate along the direction of propagation; x and y are the spatial coordinates normal to the direction of propagation; t is the temporal coordinate; c_0 is the sound velocity; ρ_0 is the density of mass; δ is the diffusivity of sound in the tissue; β is the non-linearity parameter. Under directive sound beam hypothesis, this equation allows calculation of the pressure pulse at any spatial location, given an arbitrary source pressure.

Among the currently available algorithms devoted to solve the KZK equation, the one implemented by the Bergen code has been chosen: this code works in a temporal frequency domain and in a three-dimensional space. For each frequency the code calculates the values of pressure at a fixed depth, upon a grid normal to the direction of forward propagation. The values on the first grid at $z = 0$ represent the boundary condition: in our case the boundary condition is defined by the pressure field at the probe-body interface.

The code has been modified in order to be adapted to the specific features of the simulation tool such as the frequency dependent tissue absorption and the wideband transmitted pulse.

10.3.3 A comprehensive simulation tool

The simulation tool has been conceived to work in a three-dimensional space, making it possible to evaluate the acoustic field at a given region of interest, arbitrarily located, and to assess the response produced by the scene and received at the probe of an ultrasound scanner.

In the following the structure and configuration of the simulation tool used for the investigation described in this chapter are depicted.

Transmission

The assumed probe is a linear, equally-spaced array placed on the x axis. The number of active elements and the inter-element spacing (pitch) are variables that could be freely fixed. The array composed by the active elements is centered at $x = 0$ and the ultrasound beam generated by a focused delay-and-sum beamforming is pointed perpendicularly to the array baseline, i.e., along the z axis. The transducers of the probe act as band-pass filters for the excitation signals applied to them.

The electrical waveform exciting the probe transducers can be arbitrarily set, with an emitted pressure level calculated in order to obtain the desired value at the focus point. The depth of the focus point is set at the beginning of the simulation as well as a possible apodization window used in transmission.

Moreover, the non-linear behaviors of the power amplifier of the transmission chain and of the probe transducers are taken into account. The growth of second-order non-linear components in the transmitted signal is achieved by adding a quadratic component to the linear pulse generated in the previous part. For this reason, the transmitted signal shows a level of second harmonic with respect to the fundamental that could be set by the user, according to the values expected in real equipment. Also the clipping effect due to the bridge of diodes traditionally present in the transmission chain is simulated. The main effect is the introduction of a third order non-linearity in the transmitted signal.

Finally, the simulator gives the possibility to implement the multi-pulse techniques previously discussed and the chirp coding, easily setting the number of pulses to emit and their waveforms.

Propagation and interaction with microbubbles

When an ultrasound pulse propagates through biological tissue, it is subjected to different effects. The geometrical spreading and the thermoviscous absorption have to be considered. However, before explaining how they have been introduced, the definition of the scene to be simulated (i.e., the bubble population to be analyzed) is addressed.

In the developed simulator, there is the possibility to choose the number, position and dimension of the resolution cells to consider. The user can set, inside every cell, the number of bubbles and linear point-like scatterers. The latter are used to simulate the tissue response. In this way, the tissue inside the resolution cell is assumed to have a linear response, but this does not mean that the tissue placed between the probe and the resolution cell should be considered linear. In fact, the scene is composed of a small number of resolution cells (close to each other) devoted to contain the bubble population. The ultrasound propagation between the probe and the scene can be simulated by taking into account all the tissue characteristics that are judged to be relevant.

Concerning bubbles, a random generator assigns a specific radius, chosen according to a given probability density function, to each bubble. Also the bubble's position (in the three-dimensional space) inside the resolution cell, is generated in a random way. In the same manner the position and the reflectivity factor of each linear scatterer are assigned.

The geometric attenuation is computed on the basis of the distance between each probe element and the center of each resolution cell. The thermoviscous effect attenuates the transmitted signal proportionally to the distance and the frequency. In this way, using the focused beamforming, a set of signals, opportunely attenuated and filtered at the center of each resolution cell, was computed. The sum of such signals (one for each probe transducer) produces the ultrasound pulse that excites bubbles and scatterers contained inside the considered cell. This approximation is precise enough if the dimension of the resolution cell is small (e.g., cubic cell, 1 mm side). At this point, exploiting the code related to the behavior of a single microbubble subjected to an ultrasound pulse, the echo backscattered from each bubble is calculated. Instead, the linear scatterers only reflect the impinging signal, multiplying it for their own reflectivity factors.

The non-linearity arising from the propagation of an ultrasound field inside a biological tissue can be taken into account by using the techniques described in the previous section to compute the signal impinging on the center of a resolution cell. However, such techniques are very heavy and require special attention for correct application. It was verified that when the pressure desired at the focal point is in the order of 50–150 kPa (as necessary in the case of non-destructive UCA imaging), the second harmonic component due to the tissue non-linearity is small or negligible with respect to that generated by the hardware characteristics. Therefore, although the simulation tool allows the non-linear propagation to be taken into account, typically only a preliminary assessment of its relevance is carried out, before starting a simulation session. If it is verified that the tissue contribution to the second harmonic component is not dominant, the tissue is assumed to be a linear dissipative medium.

Reception and post-processing

On the reception side, the responses generated by bubbles and scatterers are collected and summed at each element of the probe, after the same attenuation effects already observed in the transmission side (i.e., geometrical spreading and thermoviscous absorption). The non-linearity in tissue propagation is neglected due to the weak pressure level of the acoustic field backscattered by the scene.

The signals at each probe element are also filtered, as for the transmission side, to simulate the band-pass effect of the probe. After that, a focused beamforming is performed to generate the final beam signal. To emulate the dynamic focusing implemented in real equipment, the (reception) focus depth is set equal to the depth of the resolution cells that contain the bubble population to analyze. The number of the active probe elements and the apodizing window to be used can be set by the user. If multi-pulse strategies are adopted, the number of beam signals to generate and store depends on the specific technique considered.

Both the thermal noise in the reception chain and the relative time shift between two consecutive pulses due to the patient's movement (i.e., motion artefacts) are incorporated into the simulation tool. The user can choose the value of the SNR characterizing the beam signals and also the motion speed of the patient's body occurring during the transmission and reception of a pulse sequence. The thermal noise is considered as a white Gaussian noise that is added to the beam signals; the motion artefacts are introduced as time delays among the received signals due to a uniform rectilinear movement.

Regarding the post-processing phase, all the beam signals are weighted for the proper coefficients and summed together according to the selected multi-pulse technique. If Chirp CPS3 or Chirp PI are considered, the summed signal is further filtered by the related compression filter. Differently, using PI, CPS3 and CPS4 a conventional band-pass filter is adopted in order to increase the SNR: in particular, in term of central frequency and bandwidth, the filter used for PI is very similar to the compression filter used for Chirp PI (i.e., centered around the second harmonic). Instead, the filter used for CPS3 and CPS4 is very similar to the compression filter used for Chirp CPS3 (i.e., centered around the fundamental).

10.4 Evaluation metric

The metric adopted to compare the performance of the multi-pulse techniques includes the CTR and the SNR, before and after the post-processing phase (i.e., the summation of echoes and the temporal filtering, either using the matched filter or the conventional band-pass filter). The CTR is defined as the ratio between the power of the microbubble population echo and the power of the tissue echo. The two considered echoes are those present in the beam signal at the beamforming output, and are overlapped in time. Thanks to the linearity in the backward part of the simulation tool, it is possible to access these two echoes separately and to measure the CTR. The power of the thermal independent noise at the beamforming output is added to the tissue signal power. To assess performances, the CTR gain ($CTRG$) is introduced and defined as follows:

$$CTRG = \frac{CTR_{post}}{CTR_{pre}}, \tag{10.2}$$

where CTR_{post} and CTR_{pre} are the CTRs before and after the post-processing phase, respectively. The CTR_{pre} is evaluated for an emitted pulse with a unitary amplitude coefficient.

Analogously, the SNR gain ($SNRG$), denoting the robustness of a particular technique with respect to thermal noise, is introduced and defined as follows:

$$SNRG = \frac{SNR_{post}}{SNR_{pre}}, \tag{10.3}$$

where SNR_{post} is the ratio between the power of the total signal (including both the microbubbles and tissue contributions) and the power of the thermal noise after the post-processing phase. The SNR_{pre} is the same ratio, but evaluated before the post-processing phase and related to the emission of a pulse with a unitary amplitude coefficient. Both SNRs are referred to the beamforming output. The *SNRG* of a multi-pulse technique is expected to be less than 1 as the removal of the linear component (due to the summation of different pulses) causes a decrease in the signal power, while the summation of the independent noise sequences increases the noise power. Finally, even if the value of the SNR_{pre} changes due to a variation in the noise power, the *SNRG* remains constant.

The *CTRG* and the *SNRG* strictly depend on the magnitude of the non-linear contribution of the bubble population response. In turn, keeping fixed the UCA concentration, the magnitude of the non-linear contribution depends on the size of the microbubbles that compose the population and on their relative positions. As both the size and the position are random variables, the mean values and the standard deviations of the *CTRG* and the *SNRG* can be estimated only by running the simulated experiment many times.

10.5 Results

10.5.1 Setup description

The comparative analysis addressed in this chapter has been carried out by means of a set of simulated experiments, employing the simulation tool previously described.

A linear array with a central frequency of 6 MHz, a 84% fractional bandwidth and an inter-transducer spacing (pitch) of 0.245 mm was assumed. In transmission, 64 transducers, weighted by a Hamming window, were used to produce a beam focused at a depth of 3.5 cm. Concerning PI, CPS3 and CPS4 a pulse with central frequency of 4 MHz, a fractional bandwidth of 40%, and a truncated Gaussian envelope has been employed. Instead, the transmitted signal for Chirp PI and Chirp CPS3 is a chirp pulse with a raised cosine envelope and duration of 10 μs. The chirp parameters guarantee a bandwidth equal to that of the CW pulse. The transmission amplitudes were tuned in order to obtain a pressure peak (at the focus point) of 100 kPa, for an emitted pulse with an unitary amplitude coefficient. At such amplitudes, the added quadratic term, which models the hardware non-linearities, was tuned to obtain a second harmonic component 35 dB lower than the fundamental one.

Concerning ultrasound propagation in biological tissue, an absorption coefficient of 0.75 dB/cm/MHz was assumed. The region of interest (ROI) was defined as a parallelepiped $1 \times 1 \times 3$ mm^3 in size, centered at the focus point and aligned in the ultrasound beam direction (i.e., perpendicular to the array baseline). Inside the ROI, 60 microbubbles and 60 linear scatterers were randomly placed, obtaining a microbubble concentration of 20 bubbles/mm^3. Concerning the UCA modeling, the SonoVue^{TM} (Bracco, Milan, Italy) was adopted, using the values of the physical parameters provided in the literature [17].

In reception, the same 64 transducers as used in transmission were utilized, and the beamforming operation was carried out applying a Hamming window. To take into account the thermal noise, a white Gaussian noise was summed to the beamforming output to obtain an SNR equal to 60 dB before the post-processing phase. Finally, concerning the motion, a spatial shift between successive pulses of 1.5 µm was introduced that involved a tissue displacement of 0.004λ at a frequency of 4 MHz.

Each simulated experiment described in the next paragraph has been repeated 10 times. Each time the bubbles and linear scatterer positions, the bubble radii and the thermal noise have been randomly generated.

10.5.2 Analyses and comparison

In Figure 10.1a the *CTRG*, related to the five different techniques analyzed, is displayed. It can be noticed that the techniques relying on the third order non-linearity component centered on the fundamental (i.e., CPS3, CPS4 and Chirp CPS3) allow a higher *CTRG* with respect to the ones relying on the second order non-linearity component centered on the second harmonic (i.e., PI, Chirp PI). Several effects can account for this difference: first of all the second harmonic component lies near the upper cut-off frequency of the probe band and is partially filtered out while the non-linear component around the fundamental is received at the center of the probe band. Secondarily, since the tissue absorption increases with the frequency, the second harmonic propagating in the backward path is more attenuated with respect to the fundamental. Finally, it can be demonstrated [12] that PI (and similarly Chirp PI) is less robust to tissue movements in comparison with CPS3 and CPS4.

Another comparison can be made, looking at a specific multi-pulse technique, between the use of a chirp pulse and a conventional CW pulse. Considering both PI vs. Chirp PI and CPS3 vs. Chirp CPS3, an increase in CTR is observed when the chirp coding is used. This fact is coherent with the theoretical properties of the microbubbles, whose non-linear response increase with the impinging pulse length, as previously discussed.

In Figure 10.1b the *SNRG* related to each technique taken into account is displayed: a noticeable negative value (measured in dB) is reached for all the techniques employing conventional CW pulses. In fact, the signal power is decreased, due to the removal of the linear component, while the noise power is increased, due to the sum of independent noise sequences. Differently, the values of the *SNRG* related to the techniques employing a chirp pulse, though still negative, are considerably improved: this fact confirms the capacity of the chirp transmission strategy to enhance the SNR, compensating in this way the loss intrinsic to every multi-pulse technique.

From such an analysis one can deduce that, under the considered working conditions, the best choice in terms of both *CTRG* and *SNRG* is Chirp CPS3. Although this result has not a general validity (being restricted to the hypothesized conditions or to similar situations), one can conclude that

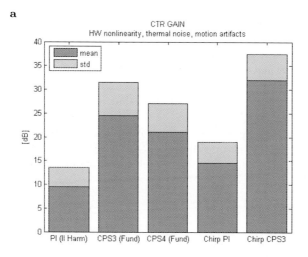

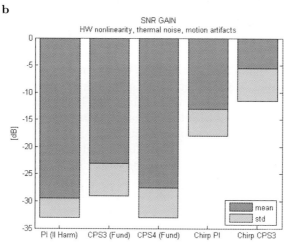

Fig. 10.1. Mean values and standard deviations of *CTRG* (**a**) and *SNRG* (**b**) regarding the multi-pulse techniques analyzed

sharp differences in the performance of the analyzed techniques surely exist. Overall, the contribution of this chapter has been to propose a method for the simulated assessment of such techniques, making it possible to compare their expected performance before choosing the technique to be adopted under specific work conditions.

10.6 Conclusions

In this chapter a comparative analysis of a set of contrast enhanced ultrasound imaging techniques has been carried out by means of simulated experiments. To this end a flexible simulation tool, able to reproduce the whole medical ultrasound chain and the interaction of ultrasound pulses with a bubble population immersed in biological tissue, has been introduced and described. The particular characteristic of this tool consists of the possibility of considering the most significant disturbing effects present in real systems. Effects such as hardware non-linear distortion of the transmitted pulse, thermal noise and body motion between the transmission of two or more pulses in temporal succession can be taken into account by the user. The setup can be easily modified on the basis of the processing technique and the desired scene that one wants to investigate.

By means of this tool the assessment of the performance of some multi-pulse techniques, employed alone or in combination with a chirp coding strategy, in terms of gain in the CTR and SNR, has been performed. The obtained results showed that sharp differences occur among the examined techniques and provided useful hints on the best technique to adopt, considering the specific working conditions addressed. In conclusion, the developed simulation tool can be considered a quite effective option for the testing of signal processing strategies related to UCAs, allowing to limit the need for expensive and time consuming in vitro or in vivo experiments.

References

1. Borsboom J, Chin CT and de Jong N (2004) Experimental Evaluation of a Nonlinear Coded Excitation Method for Contrast Imaging. Ultrasonics 42:671–675
2. Borsboom J, Chin CT, Bouakaz A, Versluis M and de Jong N (2005) Harmonic Chirp Imaging Method for Ultrasound Contrast Agent. IEEE Trans Ultrason Ferroelect Freq Contr 52:241–249
3. Chiao RY and Hao X (2005) Coded Excitation for Diagnostic Ultrasound: a System Developer's Perspective. IEEE Trans Ultrason Ferroelectr Freq Contr 52:202–216
4. Cosgrove D (2006) Ultrasound Contrast Agents: An Overview. European Journal of Radiology 60:324–330
5. Crocco M, Palmese M, Sciallero C and Trucco A (2009) A Comparative Analysis of Multi-Pulse Techniques in Contrast-Enhanced Ultrasound Medical Imaging. Ultrasonics 49:120–125
6. Crocco M, Pellegretti P, Sciallero C and Trucco A (2009) Combining Multi-Pulse Excitation and Chirp Coding in Contrast Enhanced Echographic Imaging. Measurement Science and Technology (submitted)
7. Hamilton MF, Naze Tjotta J and Tjotta S (1985) Nonlinear Effects in the Farfield of a Directive Sound Source. Journal of the Acoustical Society of America 78(1):202–216

8. Lurton X (2002) An Introduction to Underwater Acoustics: Principles and Applications. Springer Praxis Publishing, Chichester UK
9. Marmottant P, van der Meer S, Emmer M, Versluis M, de Jong N, Hilgenfeldt S and Lohse D (2005) A Model for Large Amplitude Oscillations of Coated Bubbles Accounting for Buckling and Rupture. Journal of the Acoustical Society of America 118(6):3499–3505
10. Misaridis TX and Jensen JA (2005) Use of Modulation Excitation Signals in Ultrasound. Part I Basic Concepts and Expected Benefits. IEEE Trans Ultrason Ferroelect Freq Contr 52:177–191
11. Phillips P and Gardener E (2004) Contrast-Agent Detection and Quantification. Europ Radiol Suppl 14(8):4–10
12. Phillips PJ (2001) Contrast Pulse Sequences (CPS): Imaging Nonlinear Microbubbles. 2001 IEEE Ultrasonics Symposium. Atlanta, USA 2:1739–1745
13. Skolnik M (2001) Introduction to Radar Systems 3rd edn. Mc Graw-Hill, New York
14. Stride N and Saffari N (2005) Investigating the Significance of Multiple Scattering in Ultrasound Contrast Agent Particle Populations. IEEE Transactions on Ultrasonics, Ferroelectrics, and Frequency Control 52(12):2332–2345
15. Sun Y, Kruse DE and Ferrara KW (2007) Contrast Imaging with Chirped Excitation. IEEE Trans Ultrason Ferroelect Freq Contr 54:520–529
16. Szabo TL (2004) Diagnostic Ultrasound Imaging: Inside Out. Elsevier Academic Press, Amsterdam
17. van der Meer SM, Versluis M, Lohse D, Chin CT, Bouakaz A and de Jong N (2004) The Resonance Frequency of SonVue^{TM}. 2004 IEEE Ultrasonics Symposium. Montréal, Canada, pp. 343–345

Chapter 11

Remote Measurement of the Ultrasound Contrast Agent Concentration

Marco Crocco, Claudia Sciallero, and Andrea Trucco

Abstract. Over the past decade, ultrasound contrast agents (UCA) in the form of tiny gas bubbles were introduced to improve the echographic image quality. The recent research is devoted to exploit the potentiality of new UCA, known as targeted UCA. This contrast agent is able to selectively adhere to cancer cells, as a consequence, the number of attached microbubbles that composes the UCA should be correlated with the status of the cancer, providing in this way useful medical information. In this chapter, a method to derive the microbubble volumetric concentration, working on the signals remotely acquired by means of an ultrasound scanner, is proposed. A preliminary step is devoted to separate the echoes related to microbubbles from that of surrounding tissue, identifying a suitable signal processing technique. A non-linear regression approach, based on the support vector machine, is then considered to estimate the concentration in a region of interest. The training phase is obtained extracting several significant features from simulated signals, while the signals acquired by an echographic scanner represent the test set. The estimation accuracy, obtained from in vitro experiments, seems to be sufficient in order to provide, when coupled with the binding mechanism of the targeted UCA, useful diagnostic information about the vascularization degree, and consequently, about the staging and grading of the pathology.

11.1 Introduction

This chapter is devoted to describing the progress of a specific research, aligned with the present effort of the scientific community toward the introduction of effective quantification in ultrasound medical investigation. Such research has been carried out in the context of the TAMIRUT (TArgeted MIcrobubbles and Remote Ultrasound Transduction) project whose aim is to derive the microbubble volumetric concentration working on the signals remotely acquired by means of an echographic scanner. In particular, once a 3D resolution cell is defined, the purpose is to estimate the number of microbubbles in each cubic resolution cell localized in the region of interest. The ultrasound contrast agent (UCA) developed for the project is able to adhere to cancer

Paradossi, G., Pellegretti, P., Trucco, A. (Eds.)
Ultrasound contrast agents. Targeting and processing methods for theranostics
© Springer-Verlag Italia, 2010

cells, and for this reason the concentration estimation can be useful to evaluate the staging and the grading of the tumor. This project is addressed to those pathologies, such as prostate cancer, that are difficult to diagnose, especially at the beginning, considering only the echographic image of the organ without the support of any invasive technique.

In this chapter the methodology devised and tested to estimate the bubble concentration inside a resolution cell will be described. It is worth noting that in the literature there are no works that deal with this challenging aim.

The research can be divided into two main steps. In the first a method is individuated that is able to separate the backscattered echoes produced by bubbles from that of the surrounding tissue. In the second step, the core of the research, a procedure is developed that is aimed at estimating the bubble concentration working exclusively on the received signal.

Before introducing how to evaluate the concentration of bubbles, it is important to point out the difficulties connected to this task. The echoes backscattered by two populations of bubbles with the same concentration could be very different in terms of amplitude and shape. This is due to several implicit randomness causes. One of the causes is the random position of each bubble inside the resolution cell and also the random radius of each bubble composing the population. Moreover it should be noted that it is not possible to totally eliminate the tissue contribution, and that a minimal residual of tissue echoes could vary the backscattered signal. Finally, also the backscattered echoes from bubbles in adjacent cells could cause interference.

11.1.1 General approach

Figure 11.1 shows a general scheme of the approach adopted for the above mentioned task. Supposing that we have almost eliminated the tissue contribution, through signal processing techniques that will be described in the next section, the attention is focused on the analysis of the residual signal, which is generated by the bubble population. Due to several random factors, it is impossible to devise a deterministic model for the response of a bubble population by means of which to derive the bubble concentration using a deterministic inversion process. For this reason, the concentration is considered as a random variable with a probability density function (PDF) depending on a set of features extracted from the received signal. It is also supposed that the concentration has to be chosen in a continuous range. To cope with a potential non-linear dependency of the concentration from the features extracted, a non-linear regression method has been attempted. Considering the scheme in Figure 11.1, one can notice that there are two chains in parallel: the first one is related to the processing of simulated ultrasound signals, while the other chain elaborates real signals that are derived from in vitro experiments. The first chain extract features from synthetic signals to compose a training set made up of a collection of feature vectors related to known concentrations. Since the training phase requires precise knowledge of the bubble concentration rel-

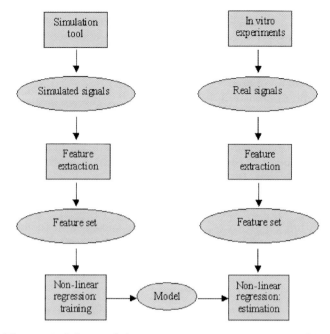

Fig. 11.1. Scheme of the concentration estimation procedure

ative to the acquired signals, it is very difficult to build up the training set through experimental data. Moreover, the assessment and tuning of the estimation algorithm may require the repetition of expensive and time-consuming experiments. To overcome these drawbacks, a simulation tool able to include the whole ultrasound chain has been exploited, allowing a totally synthetic training phase. Next, the regression model of the relationship between features and concentration, obtained in the training phase, is applied to estimate the concentration, taking as input the set of features extracted from real signals. In fact, the test set is derived from real signals, obtained from in vitro experiments.

11.1.2 Tissue rejection

To detect and separate the echoes produced by the adherent microbubbles from any other echo is a crucial task for the concentration estimation. The signal received at the echographic probe is composed of two different components: one related to the echoes produced by microbubbles, the other due to surrounding tissue echoes. To estimate the bubble concentration it is necessary, as a first step, to isolate the component due to bubble echoes from that derived by tissue.

In the literature various solutions concerning the use of multiple transmitted signals could be found [6, 9, 10]. Each pulse is transmitted with an opportune phase and amplitude coefficient in temporal succession. Combining properly the received signals it is possible to extract the second harmonic energy or to emphasize the desired non-linear order, and to suppress the linear fundamental contribution by subtraction. These techniques exploit the non-linear characteristics of the UCA that, if properly insonified, produce an echo composed of the fundamental frequency of the impinging signal and several high-order harmonics, while the tissue, if the pressure is not elevated (i.e., 50–100 kPa), has an almost linear behavior.

Among these techniques the most appropriate in the context considered in this chapter [2], able to provide the best performance in terms of contrast to tissue ratio (CTR) and signal to noise ratio (SNR), is a contrast pulse sequence (CPS) based on the transmission of three pulses [6]. CPS [0.5–1 0.5], referred to as CPS3, is a technique which transmits three identical pulses multiplied by specific amplitude coefficients, i.e., the values contained in the above vector. The sum of the received signals maintains the second and third order energy. For the application addressed here, it was decided to work only with the third order component in the fundamental bandwidth, as a consequence of the limited bandwidth of the ultrasound transducers.

11.2 Concentration estimation

The temporal signal, representing the echo of a bubble population inside a resolution cell, is the result of several random factors such as the bubbles' relative displacement, the distribution of radii, etc. Moreover, the remaining bubble parameters are not completely known and probably vary from bubble to bubble. Due to these facts, it is clearly impossible to devise a deterministic model for the bubble population response from which to derive the concentration information. Therefore, a statistical point of view has been adopted: the concentration has been considered a random variable with a PDF depending on a set of features extracted from the signal. If one knows such a PDF, one can estimate the concentration for every echo starting from the related set of features.

The problem can be split into two parts: to find a good set of features which are able to provide useful information about the concentration; to estimate the statistical dependency of the concentration from such features. This second task has been performed through a machine learning approach. A training set of features has been used to build up a function able to provide a model of the relationship between features and concentration, and consequently to estimate the concentration, given another arbitrary set of features.

11.2.1 Feature vector

The identification of significant features is particularly challenging as, because of the random factors involved, the same bubble concentration may result in very different responses. The task difficulty is further increased as the response of the bubble population, available for the estimation of the concentration, is just a portion of the original, and it is blurred by tissue echoes that are not completely rejected. Another element of disturbance is the thermal noise produced by the electronic devices.

The selection of the proper features was carried out partially through a direct analysis of the ultrasound signals and partially through feedback from the obtained results.

Features regarding both the statistical distribution of the amplitude of the signal samples and the profile of the modulus and phase of the signal spectrum were used. The use of more sophisticated features, extracted from time-frequency distributions [7] and higher order spectra (e.g., bispectra [5]), was attempted as well, but without any noticeable improvement in the concentration estimation. Also the cepstrum [1], well suited to evidence the phase information embedded in a signal modeled as a sum of multiple delayed echoes, did not provide any significant advantage.

It was verified that the signal energy is the most informative feature. However, the use of the energy can be problematic as it requires the acquisition of at least one real signal produced by a bubble population with a perfectly known concentration. This can be impossible in in vivo conditions. Moreover, the concentration estimator must work on the same conditions in which the reference signal was acquired. This means keeping the same coupling between the probe and the investigated scene: a quite difficult task. In consideration of these difficulties, the estimation of the bubble concentration discarding the echo energy from the feature vector was also investigated.

11.2.2 Non-linear regression

In general the machine learning methods deal with the construction of a function which associates, with every possible feature set, a value representing the variable to be estimated. Depending on the nature of this variable, the machine learning methods can be divided into two categories: classification methods and regression methods. In the first case the value of the variable to be estimated is chosen among a finite discrete set (the classes) without metrics among its elements; in the second case the value is chosen in a continuous and ordered set, typically the set of real numbers.

Even if the microbubble concentration inside a resolution cell is a discrete number, a regression approach was chosen since the concentration values obviously belong to an ordered set. Being $\mathbf{x}$ the n-element vector containing the feature values, and y the estimated concentration value, a regression function $f : \Re^n \to \Re$ should be derived such that:

$$y = f(\mathbf{x}), \tag{11.1}$$

where y must be as close as possible to the real concentration value related to the current set of features $\mathbf{x}$.

Since the kind of relationship between **x** and y is not known a priori, a general non-linear regression approach was chosen. Among the nonlinear regression techniques currently available, the support vector machine (SVM) [8] was adopted, since it is universally considered one of the best approaches in terms of accuracy, robustness, and computational demand.

11.2.3 Support vector machine

The SVM technique [8] reduces the non-linear regression problem to a well-known (and easier) linear one, through an implicit non-linear mapping of the features in a higher dimensional space. A limited set of parameters adequately tuned ensure a satisfactory trade-off between accuracy of the estimation, robustness and computational efficiency.

To better understand the training process, it is worth starting from the linear regression case. The linear regressor has the following form:

$$f(\mathbf{x}) = \mathbf{w} \cdot \mathbf{x} + b, \tag{11.2}$$

where **x** is the vector of the current set of features, **w** is a vector with the same dimension as the set of features, and b is a scalar value. The training process can be stated as an optimization problem with a cost function to be minimized:

$$\frac{1}{2}\|\mathbf{w}\|^2 + C\sum_{i=1}^{I}\xi_i + \tilde{\xi}_i, \tag{11.3}$$

subject to the following constraints:

$$\left.\begin{array}{r}y_i - \mathbf{w} \cdot \mathbf{x}_i - b \le \varepsilon + \xi_i \\ \mathbf{w} \cdot \mathbf{x}_i + b - y_i \le \varepsilon + \tilde{\xi}_i \\ \xi_i, \tilde{\xi}_i \ge 0\end{array}\right\}, \tag{11.4}$$

where $\mathbf{x}_i$ is the i-th set of features of the training set (containing I vectors), y_i is the related concentration value, ε is a parameter setting the accepted error between the real and the estimated concentration, C is a parameter setting the trade-off between the "flatness" (in the non-linear case the "smoothness") of the regression function, and the overall error on the estimated concentration over the training set, ξ_i and $\tilde{\xi}_i$, are slack variables (devoted to relax the constraints). The constrained minimization problem can be transformed into a dual problem, easier to solve by means of the Lagrange multipliers method. For more details, see [8].

It can be demonstrated that the optimized vector **w** can be written as a linear combination of the training vectors $\mathbf{x}_i$:

$$\mathbf{w} = \sum_{i=1}^{I}(\alpha_i - \tilde{\alpha}_i)\mathbf{x}_i, \tag{11.5}$$

where α_i and $\tilde{\alpha}_i$ are the Lagrange multipliers related to the first two inequality constraints in eq. (11.4). Since some terms $(\alpha_i - \tilde{\alpha}_i)$ can be equal to zero, $\mathbf{w}$ is a function of a training vector subset: the training vectors belonging to this subset are called support vectors. The exact number of support vectors is not fixed *a priori* but depends on the particular regression problem faced. Therefore, the complexity of the regression function, given by the number of support vectors, is a product of the optimization process. This is different from other regression methods (e.g., neural networks) where the structural complexity of the regression function has to be set as an input parameter [8].

In the non-linear case, the generic vector of features is mapped in a higher dimensional space through a non-linear vector function $\mathbf{\Phi}(\mathbf{x})$ and the optimization process is carried out in the above described manner, inside the new space. One can suppose that this mapping produces an increase of the computational burden due to the increased dimensionality of the problem. However, it can be demonstrated [8] that both the optimization process and the regression function need only the dot products $\mathbf{\Phi}(\mathbf{x}_i) \cdot \mathbf{\Phi}(\mathbf{x}_j)$ and not the explicit knowledge of the function $\mathbf{\Phi}$. Therefore, one has only to choose a dot product function $k(\mathbf{x}_i, \mathbf{x}_j) = \mathbf{\Phi}(\mathbf{x}_i) \cdot \mathbf{\Phi}(\mathbf{x}_j)$, called kernel. In this way the regression function assumes the following form:

$$f(\mathbf{x}) = \sum_{i=1}^{I} (\alpha_i - \tilde{\alpha}_i) \, k\,(\mathbf{x}, \mathbf{x}_i) + b. \qquad (11.6)$$

Note that in the non-linear case, $\mathbf{w}$ is defined only implicitly, as:

$$\mathbf{w} = \sum_{i=1}^{I} (\alpha_i - \tilde{\alpha}_i) \, \mathbf{\Phi}\,(\mathbf{x}_i). \qquad (11.7)$$

There is a wide range of functions suitable to be used as a kernel function [8]: for instance, the polynomial kernel, Dirichlet kernel, periodic kernels and B-spline kernel. Here, a Gaussian kernel has been chosen, defined as:

$$k\,(\mathbf{x}, \mathbf{x}_i) = e^{-\frac{\|\mathbf{x}-\mathbf{x}_i\|^2}{2\sigma^2}}. \qquad (11.8)$$

It can be demonstrated that the Gaussian kernel tends to produce good performance under generic smoothness assumptions, and is particularly effective when, as here, no prior knowledge about the regression function characteristics is available (further details can be found in [8]). The parameters ε, C, and the standard deviation σ of the Gaussian kernel have been chosen trying different values and retaining the ones that provided the best final results. As a constant value of ε should be set to express the accepted error, it has been decided to use the concentration values in a logarithmic scale in order to allow bigger errors on the estimation of bigger concentration values.

11.2.4 Training set generation

The importance of a simulation tool able to synthetically generate the ultra-sound response produced by many populations of bubbles with different concentrations has already been pointed out. In particular, the simulation tool is essential to collect the signals composing the training set. Clearly, such a tool should be able to simulate the signals collected by a real ultrasound scanner as a response of the propagation of a pulse (produced by the same scanner) inside the biological tissue and its interaction with a bubble population.

The simulation tool developed to this end allows us to take into account the frequency and spatial characteristics of the ultrasound probe, the non-linear behavior of the power amplifiers and the transducers used in transmission, the non-linear propagation inside the biological tissue, the level of thermal noise affecting the reception circuits, and the errors connected with the digitalization of the acquired signals. Concerning bubbles, they can be randomly located inside a predefined volume with radii chosen according to the related PDF. The bubble response to ultrasound excitation is computed by means of a refined model, as a function of the incident pressure and waveform. The response of the tissue surrounding the bubbles can be considered or neglected, according to necessity.

Further details on this simulation tool and some application examples can be found in another chapters of this book.

11.3 The experimental set-up

In this section a description of the in vitro experiments is provided. These experiments employed an ultrasound scanner equipped with an ad hoc electronic board able to recover, memorize and transfer to a standard PC the band-pass signals at the beamforming output. (The scanner and board characteristics are the subject of another chapter of this book.) In this way, it is possible to work with real data that can be opportunely processed and analyzed off-line. These real data are used to test the effectiveness of the proposed method for the concentration estimation. To this end, it was necessary to acquire the ultrasound signals backscattered from a significant number of different bubble concentrations, covering a wide range of possibilities. The simulation tool was used to collect the signals necessary to train the estimator. Obviously, the parameters of the simulation tool were tuned in order to agree as much as possible with the experimental set-up.

The considered scene is the following: an 8 mm-diameter channel embedded in a phantom simulating the acoustic properties of human tissue is positioned at a depth of 1.7 cm, equal to the focus distance of the ultrasound probe connected to the scanner (Fig. 11.2). Inside the channel, the ultrasound contrast agent flows at constant velocity, opportunely diluted in water so as to obtain the desired bubble concentration. Five different dilutions are taken into

Fig. 11.2. Experimental scene: tissue mimicking phantom and linear array probe

account, corresponding to 2, 4, 8, 16, 32 bubbles per mm^3. As mentioned previously, the multi-pulse technique used to abate the tissue response is CPS3. In Figure 11.3 the ultrasound images of the flow channel, obtained when the concentration is of 4 and 32 bubbles per mm^3, are shown.

The signals at the beamforming output are acquired and processed off-line, including the recombination of the three pulses of the CPS3 technique. For each dilution, the signals coming from inside the flow channel have been segmented into portions equivalent to a 1 mm of depth. A feature vector is extracted from each signal portion. In this way, for a given dilution, it is possible to consider a significant number of different realizations.

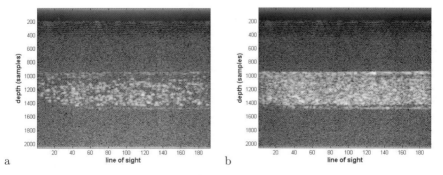

Fig. 11.3. The ultrasound images of the flow channel after the recombination of the CPS3 pulses: (**a**) concentration of 4 bubbles/mm^3; (**b**) concentration of 32 bubbles/mm^3

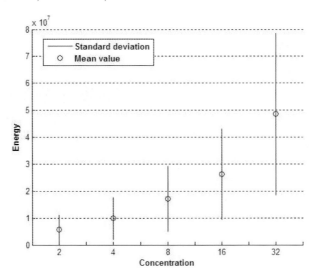

Fig. 11.4. The energy, in terms of mean value and standard deviation, for five different UCA concentrations

11.4 Results

On the basis of the analysis of the real signals, Figure 11.4 shows the energy, in terms of mean value and standard deviation as a function of the above mentioned concentrations (i.e., 2, 4, 8, 16, 32 bubbles per mm^3). The graph confirms a significant correlation between the energy and concentration, even if the standard deviation assumes noticeable values. This observation makes clearer the difficulties to overcome in estimating the UCA concentration.

Another preliminary consideration concerns the relationship between the UCA dilution and the actual number of bubbles inside a small volume (e.g., 1 mm^3). In fact, given the dilution, just the mean value of the number of bubbles inside a small volume is known, not the exact number of bubbles inside a small volume. For the above mentioned dilutions, the number of bubbles inside 1 mm^3 can be modeled as a random variable with a Poisson PDF, as shown in Figure 11.5. As a consequence, in assessing the results, it will be useful to keep in mind that the standard deviations of the estimated concentrations cannot decrease over the intrinsic standard deviations of the Poisson PDF, shown in Figure 11.5.

In the following, the results of the concentration estimation are shown for four different combinations. The first two combinations make use of the energy as an element of the feature vector. Figure 11.6 shows the results obtained on simulated (Fig. 11.6a) and real signals (Fig. 11.6b). The training set is the same in both the cases and it is composed exclusively of simulated signals. Moreover, when the test signals are simulated (Fig. 11.6a), they are different from those composing the training set.

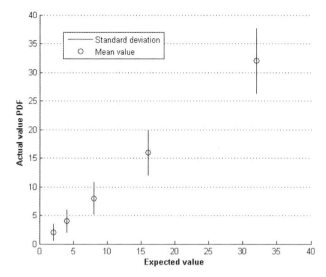

Fig. 11.5. Mean value and standard deviation of a Poisson random variable corresponding to the number of bubbles inside 1 mm^3 for five different dilutions

One can notice that a good agreement between the mean estimated value and the real mean value of the concentrations is reached, while the ratio between mean value and standard deviation is roughly constant. When real data are applied to test the estimator (Fig. 11.6b), the performance is similar to that obtained with synthetic signals. Therefore, the simulation of the responses produced by bubble populations with different concentrations is an effective option for the training of the estimator. Taking into account also the intrinsic standard deviations shown in Figure 11.5, the obtained results are considered satisfactory.

The last two combinations are referred to signals that have been normalized by their energy value. In other words, the energy information is disregarded and the energy is not an element of the feature vector. As for the previous case, Figure 11.7 shows the comparison between a test set composed of simulated signals (Fig. 11.7a) and a test set composed of real signals (Fig. 11.7b). One can notice a fair agreement between the mean estimated value and the real mean value of the concentrations, except for the highest one which results, especially for the real test set, underestimated. Unfortunately, the standard deviations are significantly increased, if compared to those obtained when the energy is retained.

11.5 Conclusions

In this chapter a method to estimate the volumetric concentration of a microbubble population surrounded by biological tissue, starting from the

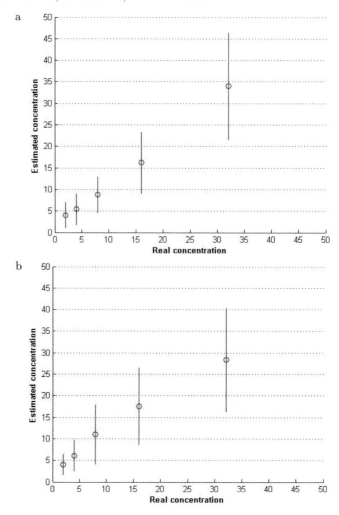

Fig. 11.6. The estimated concentration versus the real mean concentration, when the energy is retained into the feature vector. (**a**) The training set and test set are simulated. (**b**) The training set is simulated and the test set is composed of real signals

echographic signals acquired by an ultrasound medical scanner, has been described. In order to face the several random factors inherent to the problem, a statistical method based on a machine learning approach has been devised. The necessary amount of data required to train the estimation procedure has been built up by means of a suitable simulation tool, able to encompass the whole ultrasound chain behavior. Some practical tests have been conducted with a real echographic device and a microbubble population solution flowing inside a phantom reproducing the human body acoustical properties. A

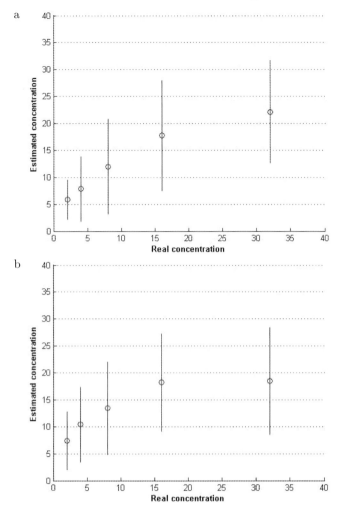

Fig. 11.7. The estimated concentration versus the real mean concentration, when the signals are normalized and the energy is discarded from the feature vector. (**a**) The training set and test set are simulated. (**b**) The training set is simulated and the test set is composed of real signals

set of echographic signals coming from different contrast media dilutions has been acquired. After a post processing phase devoted to reject the echoes coming from the phantom, a set of features has been extracted from the signal portion correspondent to each resolution cell. Starting from such features, an estimated concentration value has been derived, exploiting the regression function previously obtained from the simulated data.

Two strategies have been applied: to retain the echo energy as a significant feature or to discard this information normalizing the signals by their energy.

The first option has yielded the best performance in terms of estimation accuracy, but requires a reference signal related to a known bubble concentration. The second option avoids this requirement, allowing a quite poor estimation performance to be obtained, especially for the highest concentrations. From the preliminary tests of the estimation method proposed in this chapter, one can conclude that, if the energy is retained, the estimation accuracy could be sufficient to provide useful diagnostic information about the vascularization of the investigated region.

References

1. Childers DG, Skinner DP and Kemerait RC (1977) The Cepstrum: A Guide to Processing. Proceedings of the IEEE 65:1428–1443
2. Crocco M, Palmese M, Sciallero C and Trucco A (2009) A Comparative Analysis of Multi-Pulse Techniques in Contrast-Enhanced Ultrasound Medical Imaging. Ultrasonics 49:120–125
3. Hamilton MF, Naze Tjotta J and Tjotta S (1985) Nonlinear effects in the farfield of a directive sound source. Journal of the Acoustical Society of America 78(1):202–216
4. Marmottant P, van der Meer S, Emmer M, Versluis M, de Jong N, Hilgenfeldt S and Lohse D (2005) A model for large amplitude oscillations of coated bubbles accounting for buckling and rupture. Journal of the Acoustical Society of America 118(6):3499–3505
5. Nikias CL and Mendel JM (1993) Signal processing with higher-order spectra. IEEE Signal Processing Magazine 10:10–37
6. Phillips PJ (2001) Contrast pulse sequences (CPS): imaging nonlinear microbubbles. IEEE Ultrasonics Symposium 2:1739–1745 Atlanta, USA
7. Qian S and Chen D (1999) Joint Time-Frequency Analysis. IEEE Signal Processing Magazine 16(2):52–67
8. Scholkopf B and Smola AJ (2002) Learning with Kernels. Massachusetts Institute of Technology Press
9. Shen C-C and Li P-C (2003) Pulse-Inversion-Based Fundamental Imaging for Contrast Detection. IEEE Transactions on Ultrasonics, Ferroelectronics and Frequency Control, 50(9):1124–1133
10. Umemura S et al. (2003) Triplet Pulse Sequence for Superior Microbubbles/Tissue Contrast. IEEE Ultrasonics Symposium 429–432

Chapter 12

Bubble Behavior Testing (BBT) System for Ultrasound Contrast Agent Characterization

Francesco Guidi, Riccardo Mori, Hendrik J. Vos, and Piero Tortoli

Abstract. The acoustic characterization of Ultrasound Contrast Agents (UCA) can only be based on equipment having high sensitivity (to be able to detect the echoes produced by single microbubbles) and flexibility (to adapt to a variety of experimental conditions). In this chapter, the Bubble Behavior Testing (BBT) system is presented, and shown as an ideal tool to report on the behavior of UCA in ultrasound fields. First, its basic configuration is described (including the front-end circuits to-from two single-element transducers as well as the digital resources for transmission of arbitrary signals and processing of received echoes). Two applications of the BBT system are then discussed. The interrogation of microbubbles freely floating in a water tank, is shown to be useful to characterize the UCA by observing their response to ultrasound force. Coupling of the BBT system to a synchronized high-speed optical camera is finally demonstrated to be capable of tracking the echoes of a single deflating bubble, i.e. with variable diameter.

12.1 Introduction

The results of studies addressed to the characterization of UCA are strongly influenced by the adopted experimental conditions. To deepen into such studies it is necessary to increase the sensitivity of the equipment used and to refine the developed mathematical models.

As the analysis of bubble population doesn't allow specific properties of individual bubbles to be understood [18, 25, 26], methods dedicated to the observation of single bubbles are of utmost importance. Single bubble characterization usually requires sophisticated, complex and expensive equipment. Most of the problems come from the difficulty in isolating the target bubble whilst preserving correct bubble life conditions, removing any possible source of interference and having enough sensitivity to capture weak echo signals backscattered from the single micro-bubble.

Single bubbles are usually observed by optical methods making use of high-speed cameras [4, 14] or laser beam light scattering [15, 24]. The camera approach can be very powerful but requires that the sample volume is limited

Paradossi, G., Pellegretti, P., Trucco, A. (Eds.)
Ultrasound contrast agents. Targeting and processing methods for theranostics
© Springer-Verlag Italia, 2010

to a very small region, and that the microbubbles are therefore either fixed in some kind of gel, or into Opticell, or forced to flow in small and ultrasound (US)-transparent tubes [27]. On the other hand, the laser approach removes the volume constraint, but provides only rough indications about the instantaneous bubble diameter [15]. In both cases, the combination with acoustical equipment capable of transmitting suitable signals and receiving with sufficient sensitivity the weak backscattered echoes, is crucial.

Most diagnostic applications assume the use of resonating bubbles, therefore it is mandatory to know the conditions under which this phenomenon is maximized. Many experiments look at the energy of the echo and try to individuate the resonance condition aimed at obtaining some parameters and characteristics predicted on the basis of a model. For example, the estimation of shell parameters is usually made by fitting model prediction to the experimental results [5, 20, 29]. Unfortunately, the bubble oscillation strongly depends on the pressure amplitude (there are reported some "regimes" and "thresholds" [1, 7, 13]) and the most convenient model is fundamentally linear, resulting in low echoes, usually not easily detectable.

However, it is known that during US excitation of UCA, linear behavior is far from being effective due to different mechanisms that induce asymmetrical bubble oscillations (like in "compression-only" behavior [10, 20]) giving a significant harmonic contribution. In the case this is the object of the investigation, a wide receiver bandwidth is an important feature to be able to provide an exact description of bubble behavior.

In some cases, specific driving US pulse combinations can be useful. These points push to using a wide transmitter-transducer combination and a programmable waveform generator to provide complex excitation schemes.

Some relatively slow phenomena like bubble displacement or gas dissolution due to US required acquiring echoes for a long period of time, such seconds or tens of seconds.

A typical problem of the experiments is the difficulty to simultaneously tune some key parameters. The optical and acoustical focalization, the bubble positioning, the correct excitation scheme and echoe storing are usually a source of problems. A real-time system, capable of synchronizing the operation of acoustic and optical equipment, can help in speeding up all the critical phases.

In this paper, two different experimental set-ups, both based on the BBT system, are presented and have shown capable of acquiring detailed information about the behavior of single bubbles.

First, freely floating bubbles are excited and acoustically "imaged" through appropriate display. This approach highlights the displacement of single bubbles due to the US radiation force and the related Doppler contribution. Some other phenomena, like bubble rupture, can also be observed.

In the second set-up, an optical and acoustical combination provided information on steady bubbles observed over a long period of time. Phenomena like deflation, occurrence of resonance through more "indicators", occurrence

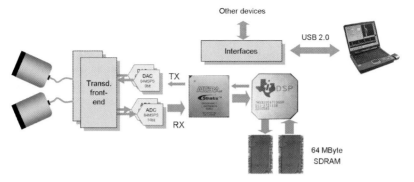

Fig. 12.1. BBT block diagram

of non-linear behavior (at medium pressure) and the characteristic response of the same bubble to different frequencies are shown.

12.2 BBT system

The Bubble Behaviour Testing (BBT) equipment is based on a single electronic board including all circuits necessary to transmit programmable signals on UCA and receive the related echoes. The board is housed in a $180 \times 200 \times 55$ mm box. In the front-end two connectors are provide to manage two single-element transducers while in the rear-end a USB 2.0 connector allows the board to be linked to a computer.

The transducer front-end is designed to be configurable, wideband, with low noise both in transmission (TX) and reception (RX), and to provide TX-RX coherent operations. The power amplifiers are designed to allow driving of a large range of transducers with different impedance and frequencies. UCA echoes, after suitable amplification and filtering, are directly converted to digital form, stored in a large memory buffer and simultaneously real-time processed and displayed on a PC monitor.

As represented in Figure 12.1, the electronics included in the board may be roughly divided into different functional sections: TX-RX front-end, storage memory and processing unit.

12.2.1 Transducer front-end

The transmitters embedded in this system are designed to drive independently two transducers with programmable sequences of arbitrary pulses produced by a powerful Arbitrary Waveform Generator (AWG).

It is possible, repeating the same pulse a programmable number of times and/or linking different types of pulses, to obtain specific pulse sequences that can be periodic, non-periodic or periodic but preceded by an appropriate transducer excitation "header".

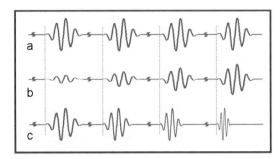

Fig. 12.2. Example of pulse sequences: (**a**) 3-cycle Hanning weighted sinusoidal pulses, (**b**) with varying amplitude (**c**) with varying frequency

In Figure 12.2 some examples are shown. The first row reproduces a weighted sinusoidal pulse, typically used for standard Doppler applications. In the second and third, pulses with varying amplitude and frequency, respectively, are shown.

The software running on the PC helps to design the desired pulse sequence through specific libraries of waveforms and functions. It is also possible to upload from a file arbitrary, waveform samples or pulse combinations.

The AWG is implemented in the Field Programmable Gate Array (FPGA) device that is described in the "Processing Unit" section. A sequence of digital samples is obtained at 64 MHz rate, converted to analog, and amplified up to 100 Vpp. The front-end is designed to drive both low and high impedance elements at frequencies ranging from 1 MHz to 16 MHz. Special care in the design of the output stage has been devoted to obtain great linearity. Harmonic studies on UCA require detecting and quantifying the harmonic contribution from microbubbles and it is mandatory to have a low harmonic contribution in the TX pulses. As an example, in Figure 12.3 the harmonic content of a transmitted pulse is shown. Here the spurious free dynamic range evaluated on the second harmonic component reach about 50 dB.

Each RX channel includes a low noise preamplifier (LNA) and a programmable gain amplifier. The latter is useful to match the dynamic range of the following 14-bit A/D converter, which operates at 64 (MSample/s).

The voltage gain ranges from 20 to 57 dB(num/V) with an equivalent input noise down to 1 nV/$\sqrt{}$ (Hz) (at 57 dB gain, 2 MHz), as shown by the black line in Figure 12.4. In the case that the transmission is performed by only one channel, the second channel can be specialized for RX-only operations. In this case, a lower equivalent input noise density down to 0.7 nV/$\sqrt{}$ (Hz) is obtained. This feature has been essential for the reception of weak signals backscattered from small microbubbles excited with pressures in the order of tens of kilopascal.

The two TX/RX channels can be arranged to work according to multiple configurations. For example, they can be simultaneously used to concurrently investigate the same region of interest (ROI) from different directions, or a

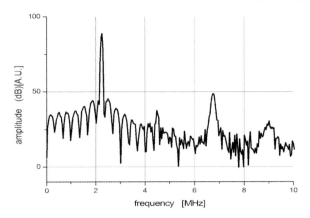

Fig. 12.3. Spectrum of a 50-cycle 2.25 MHz transmitted pulse on a 50 Ω load

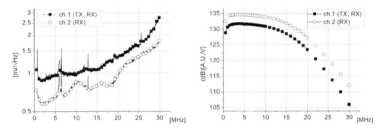

Fig. 12.4. BBT input characteristics. (Left) equivalent input noise density evaluated with a 50 Ω input load, (right) gain (output number/input voltage ratio) vs. frequency

single transducer acts as a transmitter and both transducers as receivers (see Fig. 12.5). In UCA applications the capability of using two transducers can be useful to reduce the final sample-volume and fixed echo interferences, and to implement methods to reduce phase aberration due to non-linear propagation.

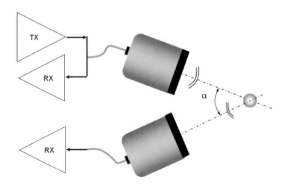

Fig. 12.5. Example of dual transducer application to single bubble studies

12.2.2 Data management

The board is equipped with a 64 Mbyte synchronous dynamic ram buffer in which input Radio Frequency (RF) samples and/or complex demodulated I/Q data can be stored. The memory acts as a circular buffer to manage a continuous acquisition process and subordinate its interruption to an external command trigger.

For each TX pulse, a programmable number (16-2048) of raw samples are stored in the memory. When the acquisition is stopped, the memory can be downloaded into a file in the host computer.

The interface manages the communication to the host computer through the USB 2.0 high-speed channel that guarantees a bandwidth of at least 20 Mbyte/s with currently available computers.

12.2.3 Processing unit

The processing unit includes two powerful digital devices: a FPGA from the Stratix family (Altera, San Jose, CA) and a Digital Signal Processor (DSP) from the TMS320C67 family (Texas Instruments, Austin, TX). This section is fully programmable and, granting a calculation power of more than 4600 Million Operations per Second (MOPS) and 600 Million Floating Point Operations per Second (MFLOPS), it is suitable to sustain intensive real-time processing of RF echo signals.

In the standard firmware implementation, the acquired RF samples are continuously sent to a digital complex demodulator followed by a programmable low pass filter, both implemented in the FPGA device. The filter is implemented in an efficient way by a chain of four-stage Cascaded Integrator Comb (CIC) filters interleaved with decimators.

The filtered samples can be stored in the memory buffer together with the original RF samples. The application specific firmware running on the DSP can access the buffer to gather demodulated data, RF data, or both, and process them according to the specific application. The DSP organizes the results of the elaboration in frames of programmable dimension that are sent to the computer through the USB 2.0 unit, for real-time display purpose.

An easy user-interface completes the system, allowing the configuration of real-time procedures, collecting data coming from the BBT system and real-time displaying or storing the elaboration results.

12.2.4 Interfaces and synchronization capabilities

The system is equipped with other input-output digital and analog channels for extended interface functionalities. An audio codec (PCM3003 from Texas Instruments) provides two input analog channels and two output analog channels that, for example, can be used configured as stereo output Doppler signal.

Synchronization capabilities are based on the management of three input/ output trigger signals to provide frequency and phase references to other instrumentations. These signals include the internal Pulse Repetition Frequency (PRF) output and a PRF input as well as the acquisition master clock input to synchronize the BBT system in a "master" or "slave" modality.

12.3 Observation of freely moving bubbles

12.3.1 Material and methods

The experimental set-up (Fig. 12.6) is based on a tank where small concentrations of microbubbles are suspended in distilled water. Two different types of microbubbles were used: BR14 (Bracco Research SA, Geneva, Switzerland), an experimental agent comprising microbubbles containing perfluorobutane, which was used at a 1:150.000 dilution, and a UCA phantom, the F04E thermoplastic microballoons containing hydrocarbon gas (C_3H_8), employed at a concentration of 10 µg/liter.

The BBT system interrogates the bubble suspension, demodulates and real-time processes the raw echo-data to extract Doppler contributions due to possible bubble movements.

Two different focused circular probes were used: a 3 MHz 70 % Fractional Bandwidth (FBW) transducer manufactured by Vermon (Tours, France), and a 6 MHz 70 % FBW by Imasonic (Besançon, France). The Vermon transducer, focused at about 80 mm, was mainly used to interrogate BR14 microbubbles while the Imasonic transducer, focused at 20 mm, was mainly used for experiments with thermoplastic bubbles.

During each experiment, the acquired echoes were stored in the BBT memory buffer, real-time processed and shown by two different displays described in the next section.

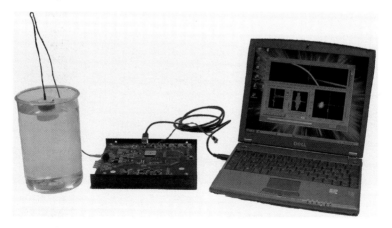

Fig. 12.6. Set-up used for studies on freely floating bubbles

12.3.2 Experimental results

The first experimental test was performed to show that the proposed acoustic setup actually could "see" individual bubbles and track the movements impressed by radiation force to each microbubble.

Figure 12.7 shows the M-mode display (i.e. the backscattered signal amplitude of subsequent echoes represented vs. depth and vs. time), obtained when the UCA concentration was so low that the US beam intercepted a few polymer F04E microbubbles. In these conditions, when the bubble population is insonified, each bubble produces a distinct echo pulse whose amplitude is converted, through the M-mode display palette, into a light spot. During subsequent Pulse Repetition Intervals (PRIs), the light-spot amplitude and position changes according to the bubble depth, producing a light-trace.

Each trace clearly shows that the corresponding bubble moves away from the transducer surface. The instantaneous velocity is proportional to the local trace slope and reaches peak values when the bubble is in the transducer focal zone.

It can be seen that brighter traces (i.e., with higher scattering properties) correspond to bubbles moving faster. This is consistent with the behavior envisaged by theory [11], which assumes that resonating bubbles are those which backscatter maximum energy and are accelerated more by radiation force [8, 12, 30].

The same phenomenon can be observed through the Multigate Spectral Doppler (MSD) display. Here, the Doppler spectra evaluated from multiple depths are simultaneously shown.

Figure 12.8 shows an example of MSD display obtained when a very few bubbles were intercepted along the beam axis. Each bubble is represented as a light spot, whose brightness corresponds to the related Doppler power, the vertical position corresponds to the bubble position along the beam axis, and the horizontal position corresponds to the bubble mean Doppler frequency.

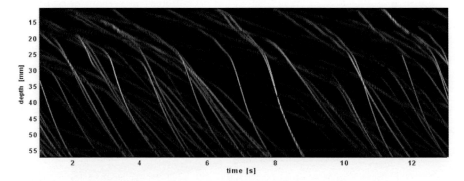

Fig. 12.7. M-mode display of F04E microspheres insonified by Imasonic transducer with 8-cycle 8 MHz pulses at PRF = 5 kHz, 1.5 MPa

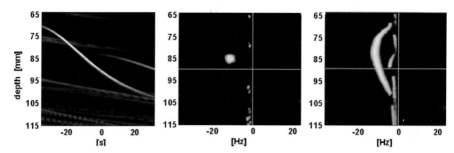

Fig. 12.8. F04e microspheres insonified by Vermon transducer with 4-cycle 4 MHz pulses at PRF = 4 kHz and 400 kPa of PNP. (Left) m-mode representation of a few bubbles, (center) MSD evaluated at t = 12 s, (right) history of the MSD display integrated in the 6–20 s time interval

On the left, a portion of M-mode display reports the trace evolution in a short time interval. Here one trace is brighter and faster than the other in the background. In the center, the corresponding MSD, evaluated around at t = 12 s, clearly shows a high intensity peak and more other weak peaks.

Due to the specific setup, the bubble displacement results roughly parallel to the beam axis, and the mean frequency can be directly converted to mean velocity through the well known Doppler equation (see for example [28]) with zero angle.

In this example F04E microbubbles have been fired with 4-cycle 4 MHz pulses repeated at 4 kHz and a 400 kPa peak negative pressure (PNP). A maximum Doppler shift of 12 Hz was measured corresponding to an average velocity of about 2.2 mm/s. However, it should be kept in mind that each bubble only actually moves during the US excitation, while remaining more or less still during the remaining part of the PRI. Considering the selected PRF and burst length, while neglecting transitory effects, a velocity of about 0.55 m/s, reached during excitation, can be estimated for this bubble.

On the right, the superimposition of many different MSD evaluated in the time interval $t = 6$–20 s shows the depth-Doppler shift pairs experimented by the bubbles during their travel in the specified time interval.

An accurate estimate of the peak velocity yielded from the radiation force can be obtained by integrating the light-spots in the MSD display over a long period of time. This integration involves a population of bubbles, surely including resonant bubbles and the long time integration should guarantee the presence of bubbles aligned at the US beam axis. The latter are expected to experience the maximum pressure and thus the maximum peak velocity, while all the other bubbles, off-axis or not resonating, move slower.

Figure 12.9a shows an integral MSD display obtained after exciting polymer F-04E microspheres for about 8 seconds. The superimposed trace corresponds to the mean velocity predicted by our model [16, 28, 30] for a resonant F-04E micro-sphere while moving along the beam axis. As expected, at each

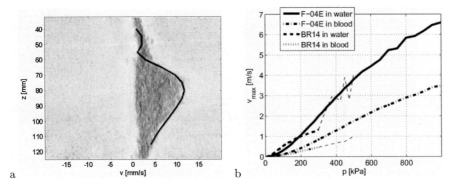

Fig. 12.9. (a) Experimental velocity profile of a population of F-04E microspheres excited by a Vermon transducer with 2.5 MHz, 10 cycle, 300 kPa pulses at 1 kHz of PRF. The superimposed black line reports the simulated resonant bubble velocities; (b) instantaneous peak velocities of resonant F-04E and BR14 bubbles excited with the same conditions as Figure 12.9a, at varying pressures

depth, the experimental measured velocities are actually distributed between zero and the maximum value predicted by the model. Around the Vermon transducer focal depth (of about 80 mm), the mean velocity of resonant bubbles is about 12 mm/s corresponding to a peak instantaneous velocity in blood of 0.9 m/s.

The agreement between the simulated and experimental results confirms the validity of the experimental assumption and the used model. This has encouraged us to estimate the peak velocity that resonating bubbles can achieve during US excitation, directly through the model. Figure 12.9b shows the peak velocities estimated for F-04E and BR14 bubbles excited with 2.5 MHz 10-cycle pulses within a range of pressures. It is clearly shown that instantaneous velocities in the range of meters/s are achieved even at pressures of a few hundred kilopascal.

12.3.3 Observation of bubble rupture

Sometimes the bubbles show a sudden change of velocity and backscatter intensity. An example is shown in Figure 12.10, in which F04E bubbles are irradiated with a 4-cycle tone burst of 4 MHz at 620 kPa. Traces with different slopes and amplitude are shown while at t = 7.5 s and 30 s two different traces suddenly stop.

It is believed that these bubbles are ruptured, or at least have suddenly changed their intrinsic properties. Changes induced by the ultrasound driving are very much reported in the literature [3, 9]. In particular, at high pressure and near the resonance frequency, bubbles experience large oscillations that can induce strong changes, depending on bubble composition. It is reported that phospholipid shelled bubbles may split into smaller bubbles maintaining

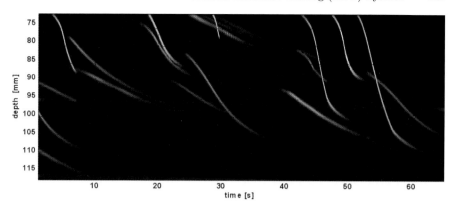

Fig. 12.10. Example of bubble "ruptures" obtained irradiating F-04E microballons with 4-cycle 4 MHz pulses at 620 kPa of PNP

the coating (processes known as "fragmentation" [1, 6, 7, 21]) while instantaneous rupture of so-called hard shelled bubbles occurs when the gas core escapes through a shell defect producing un-coated gas bubbles (known as "sonic-cracking" [4]).

Whenever a rupture event happens, the bubble changes its backscattering properties in both amplitude and phase and this involves a decorrelation between subsequent echoes.

Such phenomena are easily detectable observing the correlation between subsequent echoes through the spectrum evaluated respect to the PRF (Doppler method). Here a strong variation in the amplitude and position of the light spot is coupled to a typical wideband noise, as shown in the right side of Figure 12.11.

The M-mode display on the left shows, at around t = 7.3 s, a zoomed version of the trace slope change, obtained exciting BR14 bubbles by 2 MHz

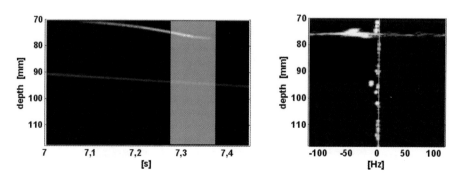

Fig. 12.11. Example of bubble rupture simultaneously shown through M-mode (left) and MSD (right) display obtained irradiating BR14 bubbles with 2-cycle 2 MHz pulses at 1 kHz of PRF and 500 kPa of PNP

pulses. The highlighted portion is then considered to produce the MSD frame shown on the right.

12.4 Observation of steady bubbles

12.4.1 Materials and methods

The experimental setup was based on the combination of acoustical and optical systems, as roughly shown in Figure 12.12.

An optical and acoustically transparent small fibre (150 μm inner diameter), immersed in a water tank, was used to hold the microbubbles under test (BR14 (Bracco Research SA, Geneva, Switzerland) and Definity (Medical Imaging, North Billerica, MA, USA).

Two focussed single element transducers (PA076, Precision Acoustics, Dorchester, UK; 1-inch C381-SU, Panametrics-NDT, Waltham, MA, USA) were co-focally positioned in such a way to intercept a portion of the fibre. Special care has been dedicated to avoid possible fixed echoes from the fibre walls.

The same portion of the fibre, i.e. the ROI, was illuminated by a continuous light source and placed on the optical plane of a long working distance (LUM-PlanFL 40 x, N.A. 0.8/W, Olympus, Tokyo, Japan) objective. The objective was mounted on a microscope (Olympus BX-FM, 2X/4X extra zoom), which projects the images on a commercial digital camera (Redlake, MotionProTM, San Diego, CA, USA), capable of collecting up to 10000 frame/s, and having 4 GB circular memory storage capability. The final resolution was up to 13.4 pixels/μm.

The transducers were connected to the BBT system, configured in such a way as to use one transducer as a transmitter and the second as a receiver. Frame synchronization to the BBT system was applied in order to maintain

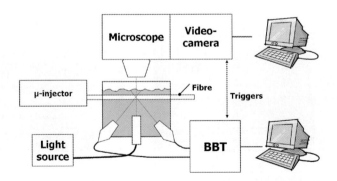

Fig. 12.12. Setup used for studies of steady bubbles. Two US transducers and the microscope are co-focused on a single point into the fibre. BBT and video-camera systems opportunely synchronized store the acoustical and video data on PC files

the correlation between each frame and the US echo received at the corresponding PRI.

Both video and US sub-systems were controlled by real-time software to
manage all the critical parameters and to show the acquired signals.

The amount of memory available for both the BBT system and the camera
was adequate to record data over intervals of several seconds.

Before any acquisition, the target bubble was carefully positioned in the
focal zone taking care that no other bubbles were close, to reduce as much
as possible any source of interference. Each bubble was excited interleaving
pulses at 5 different frequencies to be able to obtain, from the same bubble,
the response to different excitation conditions.

Each pulse, fired at low PRF (250 Hz) to avoid any possible memory
effects, was long enough (15 cycles) to allow neglecting transitory effects, and
at low pressure (50–130 kPa).

For each TX pulse, the full echo and a single frame were stored. The
entire video frame and corresponding backscattered echoes were then analyzed
through Matlab (Mathworks Inc., Natick, MA, USA) to extract the bubble
radius, the first and the second harmonic amplitudes (computed by narrow
band integration in the FFT power spectrum), as well as the instantaneous
relative phase.

12.4.2 Experimental results

The observation of single bubbles insonified as described above, allowed a
detailed study of the deflation phenomenon [3,22]. The bubbles in fact dissolve
very slowly when driven at low pressure, and usually reach a stable minimum
dimension still producing detectable echoes. At higher pressures, they dissolve
faster (in tens of pulses), and sometimes the rupture of the shell into detectable
fragments or shell modifications can be observed, as shown in Figure 12.13
[10, 20].

In Figure 12.14, we report the first and second harmonic amplitude of
echo, the relative echo phase and the radius of a deflating bubble driven at
medium-pressure amplitude (130 kPa). As an example, we report also the raw
echo signals, corresponding to different phases of the deflation process.

The bubble deflates slowly, until it becomes smaller than 1.7 µm; here the
size suddenly decreases much faster and in a few PRIs it reaches a stable state
at about 1.2 µm. The amplitude is stable during the first phase, showing a
peak immediately before the start of the fast shrinking process. Only here a
strong second harmonic echo component is detectable. The phase changes by
more than $\pi/2$, mainly during fast shrinkage. The backscattered signals reflect
this behavior. Particularly the echo in correspondence to the amplitude peak
shows a visible non-linear oscillation. The last, smaller, echo is also indicating
strong non-linearity, but with a different shape due to a different phase of
second harmonic.

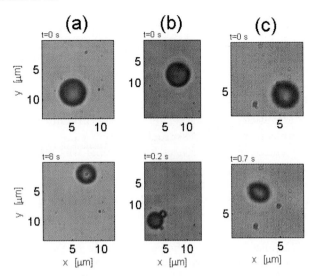

Fig. 12.13. Examples of phenomena observed during ultrasound induced deflation. (**a**) normal radius reduction, (**b**) formation of smaller bubbles maintaining consistent shell structure [22, 23], (**c**) initially spherical to a non-spherical shape

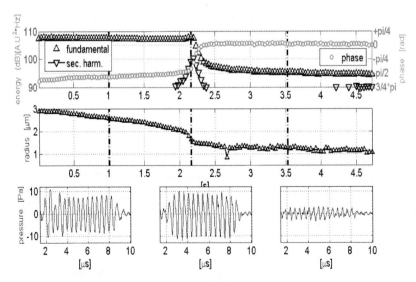

Fig. 12.14. Deflation of a Definity bubble induced by medium pressure driving (15-cycle 2.5 MHz pulses at 250 Hz of PRF and 130 kPa of PNP). (Top) First and second harmonic amplitudes and the (relative) phase displacement vs. the excitation time; (center) radius curve; (bottom, from left to right) echo signals at 1.0 s, 2.2 s, 3.5 s

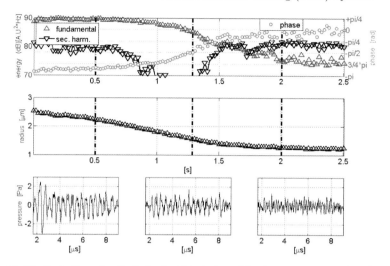

Fig. 12.15. Deflation of a Definity bubble induced by low-pressure pulses (15-cycle 2.5 MHz pulses at 250 Hz of PRF and 70 kPa of PNP). (Top) First and second harmonic amplitudes and the (relative) phase displacement vs. the excitation time; (center) radius curve; (bottom, from left to right) echo signals at 0.5 s, 1.3 s, 2.0 s

The same experiment was then carried out using US pulses with lower pressure. In Figure 12.15 the amplitude, phase and radius of a deflating bubble driven at 2.5 MHz, 70 kPa is reported. The bubble decreased in size following a typical sigmoid curve [3] and reaching a stable value. When the deflation starts, the amplitude initially keeps a stable value and subsequently shows a decrease. Instead of the amplitude peak shown in Figure 12.14, here a hump is shown. The second harmonic amplitude show higher values in the first and in the last deflation phases, while no peak is detectable during the faster shrinking phase. The (relative) phase displacement overcomes $\pi/2$ in the overall excursion.

The results indicate that: i) the deflation process is characterized by an abrupt shrinkage when driven by medium pressure involving dramatic change in the mechanical properties while with lower pressure excitation a more gradual dissolution is observed as evidenced by the radius curves; ii) correspondingly the amplitude shows a clear peak with medium pressure and a smoothed knee with lower pressure, while maintaining high scattering properties as evidenced by the delayed decrease of the amplitude curve if compared with the radius curve, iii) the relative phase lag covering an overall excursion higher than $\pi/2$ indicates that the bubbles pass in both cases from a dimension higher to a lower than resonance size; iv) the second harmonic content is related to an oscillation in a non-elastic manner; we expect that in the first case it is related to strong oscillation in correspondence to the amplitude peak while in the second case to compression-only behavior in the final stage [2, 20].

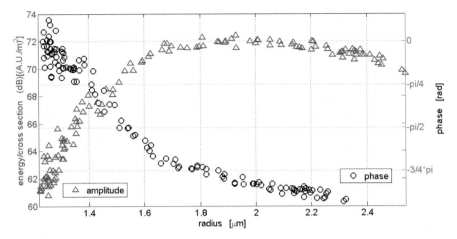

Fig. 12.16. First harmonic normalized amplitude and (relative) phase displacement of a Definity bubble excited by 15-cycle 2.5 MHz pulses at 250 Hz of PRF and 70 kPa of PNP

Figure 12.16 reports the integrated power of the first harmonic normalized to the geometrical cross section and the (relative) echo phase vs. radius. The latter is represented assuming a 0 phase when the bubble assume the minimum radius (around 1.2 μm). The resonance condition can be defined in terms of maximum scattered power or in terms of phase displacement for the bubble as a linear oscillator [18]. Here maximum scattering appears at about 1.9 μm, while a phase lag of $\pi/2$ corresponds to about 1.6 μm. The difference between the two values is an indication of the damping [19].

These curves summarize the scattering properties with respect to the bubble size disregarding the deflation phenomena itself, that could play an important role concerning correct data interpretation.

A second factor of uncertainty in the interpretation of results comes from the assumption of 0 lag at the minimum radius. Is the bubble really much smaller than the resonant radius? The bubble could be so damp as to have no lag at a very small radius, and resonate in terms of phase, very far from the resonance in terms of scattered energy.

Both the uncertainties can be partially reduced observing the characteristics of many bubbles, with different histories as different initial and final radii. We observed that bubbles showed in the relative scattered energy curves the maximum amplitude at about the same size and correspondingly a phase displacement with the same trend.

12.5 Conclusions

This paper has presented an advanced electronic (BBT) system dedicated to UCA characterization. The main features of such a system are flexibility and sensitivity. A programmable TX can excite one or two transducers with arbitrary sequences of arbitrary waveforms, while an ultra low-noise receiver can detect and acquire the low amplitude RF echoes backscattered from single microbubbles even with small and not resonating bubbles.

The BBT system has been used in a stand-alone configuration to characterize UCA by observing their reaction to radiation force. Coupling (and synchronization) with a high-speed camera has allowed a detailed investigation of the behavior of single microbubble deflation due to prolonged exposure to low-pressure US pulses.

References

1. Bloch SH, Wan M, Dayton PA and Ferrara KA (2004) Optical observation of lipid- and polymer-shelled ultrasound microbubble contrast agents. Appl Phys Lett 84(4):631–633
2. Borden MA et al. (2004) Surface behavior and microstructure of lipid/ PEG-emulsifier monolayer-coated microbubbles. Colloids Surf B 35:209–223
3. Borden MA et al. (2005) Influence of lipid shell physicochemical properties on ultrasound-induced microbubble destruction. IEEE Trans Ultrason Ferroelectr Freq Control 52(11):1992–2002
4. Bouakaz A, Versluis M and de Jong N (2005) High-speed optical observations of contrast agent destruction. Ultrasound Med Biol 31(3):391–399
5. Chatterjee D and Sarkar K (2005) A newtonian rehological model for the interface of microbubble contrast agents. Ultrasound Med Biol 29(12):1749–1757
6. Chomas JE et al. (2001) Mechanism of contrast agent destruction., IEEE Trans Ultrason Ferroel Freq Contr 48(1):232–248
7. Chomas JE et al. (2001) Threshold of fragmentation for ultrasonic contrast agents. J Biomed Opt 6(2):141–150
8. Dayton PA, Allen JS and Ferrara KW (2002) The magnitude of radiation force on ultrasound contrast agents. J Acoust Soc Am 112(5):2183–2192
9. Dayton PA et al. (1999) Optical and acoustical observation of the effects of ultrasound on contrast agents. IEEE Trans Ultrason Ferroelectr Freq Contr 46(1):220–232
10. de Jong N et al. (2007) "Compression-only behavior" of phospholipid-coated contrast bubbles. Ultrasound Med Biol 33:653–656
11. Doinikov A (1998) Acoustic radiation force on a bubble: Viscous and thermal. J Acoust Soc Am 103(1):143–147
12. Eller A (1968) Force on a bubble in a standing acoustic wave. J Acoust Soc Am 43(1):170–171
13. Emmer M et al. (2007) The onset of microbubble vibration. Ultrasound Med Biol 33(6):941–949
14. Garbin V et al. (2007) Changes in microbubble dynamics near a boundary revealed by combined optical micromanipulation and high speed imaging. J Appl Phys L 90(11):114103
15. Guan J and Matula TJ (2004) Using light scattering to measure the response of individual ultrasound contrast microbubbles subjected to pulsed ultrasound in vitro. J Acoust Soc Am 116:2832–2842

16. Guidi F et al. (2005) Acoustical imaging of individual microbubbles. Acoust Imag 28:257–265.
17. Khismatullin DB (2004) Resonance frequency of microbubbles: effect of viscosity. J Acoust Soc Am 116(3):1463–1473
18. Khismatullin DB and Nadim A (2002) Radial oscillation of ancapsulated microbubbles in viscoelastic liquids. Phys Fluids 14:3534–3557
19. Leighton TG (1994) The acoustic bubble. Academic Press, London
20. Marmottant P, van der Meer SM, Emmer M, Versluis M, de Jong N, Hilgenfeldt S and Lohse D (2005) A model for large amplitude oscillations of coated bubbles accounting for buckling and rupture. J Acoust Soc Am 118(6):3499–3505
21. Postema M et al. (2004) Ultrasound-induced encapsulated microbubble phenomena. Ultrasound Med Biol 30(6):827–840
22. Pu G, Borden MA and Longo ML (2006) Collapse and shedding transition in binary lipid monolayer. Langmuir (22):2993–2999
23. Pu G, Longo Marjorie L and Borden MA (2005) Effects of microstructure on molecular oxygen permeation through condensed phospolipid monolayer. J Am Chem Soc 127:6524–6525
24. Qiu H-H and Hsu CT (2004) The impact of high order refraction on optical microbubble sizing in multiphase flows. Exper Fluids 36:100–107
25. Sboros V et al. (2004) An in vitro study of a microbubble contrast agent. Phys Med Biol 49:159–173
26. Stride E and Saffari N (2003) Microbubble ultrasound contrast agents: a review. Proc Inst Mech Eng Proc Part H 217(H6):429–447
27. Sun Y et al. (2006) High-frequency dynamics of ultrasound contrast agents. IEEE Trans Ultrason Ferroelect Freq Control 52(11):1981–1991
28. Tortoli P, Pratesi M and Michelassi V (2000) Doppler spectra from contrast agents crossing an ultrasound field. IEEE Trans Ultrason Ferroelectr Freq Contr 47(3):716–726
29. van der Meer SM et al. (2007) Microbubble spectroscopy of ultrasound contrast agents. J Acoust Soc Am 121(1):648–656
30. Vos HJ et al. (2007) Method for microbubble characterization using primary radiation force. IEEE Trans Ultrason Ferroelectr Freq Contr 54(7):1333–1345

Chapter 13

A New Flexible Ultrasound Scanner Suitable for Research Topics

Alessandro Nencioni and Paolo Pellegretti

Abstract. The synergic use of new ultrasound contrast media (UCM) with improved and tuned ultrasound scanners (USs) implementing refined contrast processing techniques would represent the most exciting future in medical ultrasound. Particularly, the pharmaceutical industry has under advanced stage of development new targeted UCM tuned to deal with specific applications (e.g., prostate cancer diagnosis) which potentially could significantly broaden the market of ultrasound based medical examinations.

It is so clear the great effort of ultrasound scanner manufactures to develop even more powerful USs able both to satisfy the sonification requirements of new UCM and to properly process their specific acoustic signatures.

The aim of this chapter is to give an overview of the present experience on developing a new powerful US platform of Esaote which is one of the major US manufacturing companies in the UCM processing field.

13.1 An industrial experience on UCM processing

How much Esaote trusts in the potentialities of UCM is witnessed by the great effort devoted in developing dedicated US solutions since the beginning of research in this field (i.e., mid '90).

In 1995, in the real first pioneering phase of the use of contrast media for ultrasound applications, Esaote started a joint research cooperation with Bracco Research SA (Geneva – Suisse).

In 1997 Esaote presented the first preliminary result of this cooperation, that is, a dedicated version of its AU5 scanner, named, AU5 Harmonic. At that time, this unit represented one of the first commercially available offers for US scanners with UCM processing capabilities. This unit was equipped with dedicated hardware to deal with first generation UCM. Very rough processing, limited by the capabilities to be a fully analog unit was carried out. Mainly the bubble detection was carried out operating at a high level of mechanical indices exploiting the strong echoes produced by bubble destruction. Of course, the

Paradossi, G., Pellegretti, P., Trucco, A. (Eds.)
Ultrasound contrast agents. Targeting and processing methods for theranostics
© Springer-Verlag Italia, 2010

system was only able to operate in intermitted imaging mode and no capillary circulation was detectable.

In 1999 the first of Esaote's fully-digital US, the Technos, was presented. Even though very powerful in terms of processing capabilities, this first release was not significantly better than AU5 as for the UCM processing capabilities. Again, the only available UCM processing was a multipulse technique aimed at detecting bubble destruction as a consequence of high mechanical index (MI) sonifications.

The performances ensured by this unit were very good with the only available UCM on the market at that time, that is, the Levovist by Shering.

In 2000 Esaote presented the Megas Esatune, the first proposal of low level US with UCM capabilities available on the market. It was able to operate real-time at low mechanical index exploiting the strong 2nd harmonic response of 2nd generation UCMs (i.e, Sonovue, Definity, Optison, etc.). The spatial resolution and the dynamics of signal representation were quite poor and the system needed to use dedicated probes (with limited general imaging performances).

In 2001 Esaote presented a fully revised version of its top-class US Technos names Technos MPX. Basically the processing methodologies were similar to those implemented on the Megas Esatune, improved in terms of sensitivity and spatial resolution by exploiting the most sophisticated front-end and the extended processing characteritics.

The rest is recent history: in 2005 the MyLab 70 was presented, a fully new scanner with similar UCM processing characteristic as the old Technos MPX and, more recently, the top-class scanner MyLab 70 XVG.

13.2 The new scanner

In early 2002, Esaote's decided to start with an ambitious project, that is, the design of a new US with state of the art characteristics, in order to fully satisfy the processing requirements of 2nd generation UCM.

The main particularities of the new scanner are summarized hereinafter:

- 192 fully independent transmission (TX) / reception (RX) channels;
- fully linear TX/RX front-end with possibility to drive probes with arbitrary waveforms. Analog/digital apodization available both in TX and RX;
- high-dynamic receiving processing chain operating at an equivalent resolution of 20 bits;
- optional printed circuit board (PCB) with extended capabilities to digitally process radio-frequency (RF) data;
- PC based architecture with dedicated echographic section connected through a PCI interface to the host PC;
- 4 input probe connectors;

- support of a wide range of electronic probes (convex, linear, phased-array and pencil) ensuring multidisciplinary capabilities;
- extended operative range from 1 MHz up to the limit of 18 MHz;
- multimodality functionality [6], that is, B-Mode, M-Mode, color flow mapping (CFM), pulsed wave (PW) and continuous wave (CW) Doppler. Combined Duplex and Triplex functionalities are supported too;
- extended B_Mode field of view (panoramic view) very useful to have a whole image of the patient anatomy. Particularly addressed to musculoskeletal and vascular applications;
- real-time data-fusion between the US images with 3D computerized tomography (CT) and magnetic resonance (MR) volumes acquired with high-spectral and contrast resolution;
- 3D and 4D capabilities;
- extended processing based on RF data to make a quantitative evaluation of intima thickness and arterial stiffness.

In the original design a great attempt was made to ensure the best performance in terms of sensitivity, noise and TX/RX linearity of the analog front-end (all of these, mandatory particuliarities to deal with contrast media processing). Nevertheless, even though a very powerful backdoor has been kept to implement new and powerful real-time processing on the optional PCB of the previous point 4 (described in detail in a later section) no new particular processing, inherently different with respect to our previous scanners, was planned at that time.

The occasion to push ahead in this sense arose in 2004 when Esaote joined the Tamirut project.

13.2.1 General overview

This project developed as a full revision of the present top class scanner of Esaote, that is, the MyLab70.

For the sake of continuity and to underlie the strong connection with the previous unit, the commercial name of the new scanner will be MyLab70 XVG. Even though the external appearance of the two units (Fig. 13.1) is very similar, except for the presence on the new unit of the service display just above the main keyboard and for the presence of a 4th probe connector, the electronics and the performance of the new scanner are totally new and it shares only few components with its "young" brother.

A block scheme of the system is given in Figures 13.3a and 13.3b. As already stated in the previous section, the system architecture is PC based, that is, a PC is used to run both the user-interface and the main control software. The echographic electronic is connected to the PC via the Back-End Link Control (BLC) PCB board which implements a standard 32 bits PCI link (an upgrade to 64 bits PCI is in progress). This solution would allow to have the apparatus the typical powerful network connectivity and the flexible interfacing with external peripherals of PCs.

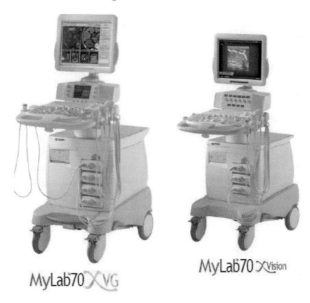

Fig. 13.1. MyLab 70XVG and MyLab70

The echographic unit PCBs are connected together with a dedicated bus named RT Bus (acronym for Real-Time Bus). This link is very similar in principle to a 32 bits SCSI bus. Every board can send/receive to any other messages structured with a header of two words defining the message source, the message address, the message type, etc. The message body follows for a total message length of up to 512 words (longer commands spanning over multiple messages are allowed). On the BLC PCB there is a master arbiter which manages the messages sent across the bus. The data rate is of 32 Mwords/s.

Fig. 13.2. Project MyLab 70XVG: echographic rack detail

As system's philosophy, all the digital links, inside the boards and between the boards are based on low-voltage differential signaling (LVDS) communication technology. Dedicated links are used where required, e.g., the RF data transfer from Digital Image Processing (DIP) to Digital Radio-Frequency Processing (DRP) boards and the RX focalized data from the Input TX-RX (ITR) block to the DIP board. LVDS is a very effective technology which allows the creation of fast links with very low power consumption. Moreover, the differential architecture ensures very good levels of noise immunity, and hence, to easily manage the routing on high density electronic PCBs like the one used in the MyLab 70 XVG project. The LVDS channels are double data rate (DDR) with a 100 MHz clock. The system master-clock is generated on the Input CW Clock (ICC) board and distributed to the whole system.

Following is a brief description of the main system components.

Input Connector Switch board

The probe interface is on the Input Connector Switch (ICS) PCB and it is realized with 4 AMP, 260 pins each, input connectors. The pin-out is maintained compatible with that used by previous USs by Esaote. A few more controls have been added (4 pins) to implement a serial communication link toward the probe. This last feature could be useful to control the electronic integrated onboard future probes (e.g., capacitive Micromachined Ultrasonic Transducer, cMUT, transducers).

Input TX-RX board

The core of the system front-end is represented by the block of 16 Input TX-RX (ITR) PCBs, which accomplishes the following functions:

- TX/RX focalization processes;
- TX driving;
- RX A/D conversion and amplifying.

Each PCB manages 16 TX/RX channels, so the system can drive independently probes with up to 192 elements. This feature, expensive from the architectural point of view, ensures the maximum driving flexibility (i.e, trapezoid view, effective support of 1.5D and 1.75D transducers [6], support of bi-planar transducers, etc.).

TX waveforms are, for transfer speed reasons of the TX data from the PC to the system front-end, shared by a group of 4 neighbor elements. The maximum waveform length is of 512 samples, 10 bits per sample. The D/A conversion is carried out by 10 bits/100 MHz converters. It is forseen the possibility to enable the interpolation of neighbor real samples to increase the waveform length to 1024 samples at an equivalent sample rate of 50 MHz. In real-time, i.e. at each pulse transmission, up to three different waveforms per channel can be managed and switched. Moreover, it is always possible to

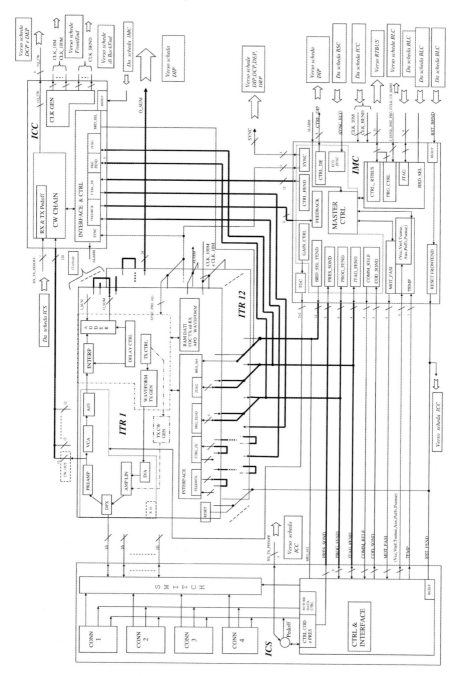

Fig. 13.3(a). Project MyLab 70XVG: block diagram

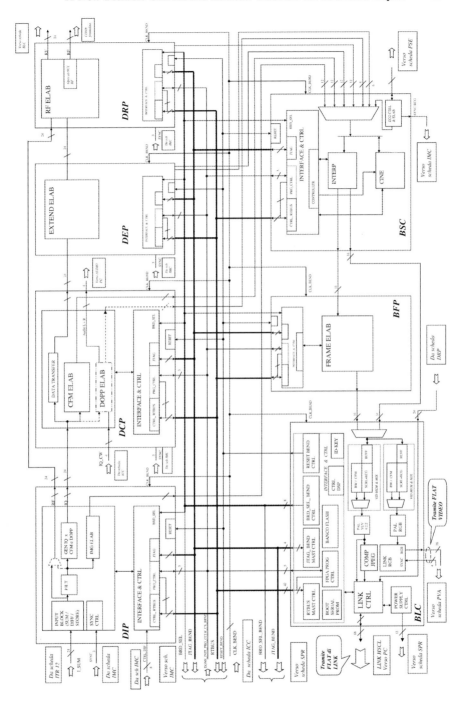

Fig. 13.3(b). Project MyLab 70XVG: block diagram

force the TX of inverted copies of each TX waveform by a dedicated inversion control of the final TX amplifier.

The RX focalization process is carried out on the base of a fixed sampling frequency of 50 MHz. To ensure the required temporal resolution to correctly focalize also high frequency probes, a temporal resolution of 10 ns is obtained increasing the sampling frequency by interpolation of adjacent time samples.

For each RX channel there is the possibility to temporarily memorize the acquired data until the reception of the next pulse echoes. This feature can be exploited to focalize the same set of acquired data in different ways (e.g., to implement the double or the quadruple receiving lines process [6]).

For each RX channel, the A/D process is based on a 12 bit converter. This means that, thanks to the system gain achieved with the coherent sum of the contributes of the echoes received by the probe elements used for the focalization process, is possible to obtain a data resolution up to the limit of 20 bits. To be able to really exploit such a high potential RX dynamic (120 dB), countermeasures had been adopted like the use of fully differential paths up to the converter input and very low noise components for the RX chain.

The RX chain is based on a first preamplifier stage with fixed gain aiming only to match the probe/cable impedance followed by an integrated variable gain amplifier (regulation range of about 44 dB) used to compensate the different probe efficiencies and the tissue/beam attenuation.

Finally, to perform TX and RX beamforming with low side lobes, proper apodization profiles can be applied both in transmission (digital apodization) and in receiving (analog apodization).

Input CW Clock board

The Input CW Clock (ICC) board generates the system's master clock of 100 MHz.

Another important feature in charge of this board is the RX focalization for CW Doppler [6]. Due to the particular characteristic of this modality, which is unique among the implemented modalities on a US, to have active at the same time both the transmission and the receiving processes, the receiving dynamic ensured by the present A/D converters (for the other imaging modalities a state-of-art converter with 12 bits of resolution and 65 MHz of maximum sampling frequency is used) is absolutely insufficient to linearly convert both the TX and the RX signals. In fact, the above mentioned converter offers a dynamic of 72 dB per channel, whereas to meet the aforesaid requirements at least 100–120 dB would be necessary. A no linear conversion would results in distortion effects which blur the received signals spectrum making it unusable. Thus, the RX focalization process of the CW is carried out by analog time-delay chains.

In our scanner, up to 60 RX channels can be used (specifically, the outmost of the active probe's aperture; the innermost being used for the transmission).

Input Master Control board

The Input Master Control (IMC) board is aimed at controlling the real-time operation of the unit. It communicates via a serial link with the ICS and by means of an LVDS link with the ITR boards (i.e., to upload focalization data, transmission waveforms, RX chain controls, etc.).

Back-End Link Control board

The main task of the Back-End Link Control (BLC) board is to provide the PCI link to the PC. On board a JPEG compressor is present to allow real-time recording of acquired images on the PC's hard-disk drive.

Digital Image Processing board

The Digital Image Processing (DIP) board is devoted to accomplish the basic image processing of received data (i.e., the B-mode and M-mode). The structure of the DIP board is shown in Figure 13.4.

It receives in input from the ITR boards the focalized data stream through a 10 bit/100 MHz DDR LVDS channel.

On the basis of the system's setting:

- one or two 20 bit, 50 Mwords/s received lines per pulse repetition frequency (PRF) (in single and double receiving lines/firing modes respectively);
- four 20 bit, 25 Mwords/s received lines per PRF (in four receiving lines/firing mode) can be processed in real time.

The output channels are:

- one 10 bit/100 MHz DDR LVDS channel to the DRP board (this link is used to transfer RF data);
- a few other LVDS channels to transfer data to Digital Color Processing (DCP) board and Digital Extended Processing (DEP) board.

RF input data are filtered by means of two banks of up to 128 FIR filters which can be switched in real time along the echoes' receiving. This feature allows a filtering of echoes coming from tissues at different depths adapted to the point to point frequency spectra changes to be made. RF filters are accomplished by two Xilinx xc2v1000-4, each one implementing a 32 multiplier and adder architecture (20 bit data, 12 bit coefficients, 38 bit resolution, with a final shift to shrink to the output resolution of 20 bits). RF filters can be configured either in serial or cascade configurations. Moreover, feedback data-paths are foreseen to implement symmetric and anti-symmetric filtering architectures (odd or even).

At the input level, two static RAM memories of 128 k * 36 bit each allow up to 8 RF scan lines for combined signal processing facilities to be memorized.

Data are baseband demodulated and logarithmically compressed to 8 bits (video resolution) using a proper look-up table (on-line, 4 look-up tables available). Output data consequent to successive sonifications, carried out for the various TX focuses of each scan line, are merged. Before being sent to the scan-converter board, data are further processed by an IIR filter which accomplishes a frame-to-frame filtering. Furthermore, to improve the scan speed, the density of scan lines can be reduced enabling the interpolation of neighbor lines to reconstruct the not actually acquired lines.

Digital Elaboration Processing board

This Digital Extended Processing (DEP) board implements complex context processing filters on processed black and white (B/W) images. These filters are aimed at making a noise reduction and to improve the overall image quality.

Digital Color-Flow Processing board

The Digital Color-Flow Processing (DCP) board is responsible for carrying out the CFM and the PW Doppler analysis [6].

Back-End Scan Converter board

The Back-End Scan Converter (BSC) board receives data processed by the other system's boards (e.g., DIP, DEP and DCP) and, driven by the system's main software, real-time accomplishes the correct scan geometry reconstruction.

Digital Radio-Frequency Processing board

The Digital Radio-Frequency Processing (DRP) board is the extension we foresaw to accomplish advanced processing of received data.

When we started the project we left empty this bus slot paying attention only to provide input and output routings to fulfill any possible future processing needing. Therefore, the input is selectable either from the DIP (RF received beamformed data just before any processing) or from the DCP. In both cases, 10 bit LVDS DDR links at 100 MHz are used. The outputs are processed data toward the BSC board or a Gigabit Ethernet link to transfer RF data to an external PC.

Participation in the Tamirut project pushed to start with a design of a board to implement a few selected algorithms among those developed by the other consortium partners.

For the sake of generality two main guidelines have been followed in the PCB design:

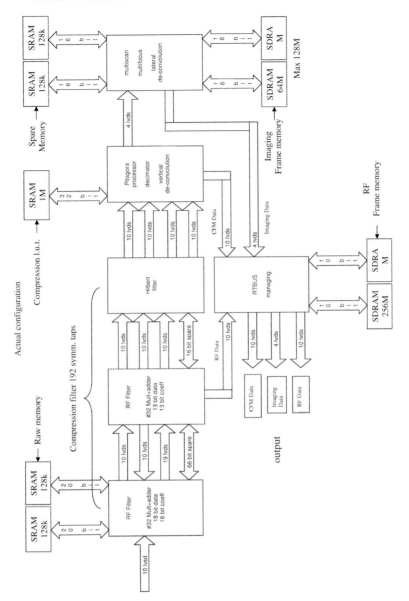

Fig. 13.4. DIP (digital image processing) board: block diagram

- developing the most powerful and general purpose hardware compatible with our US and the technological constraints (i.e., available space, power consumption and heat dissipation capabilites);
- foreseeing the possibility of real-time transfer of the RF beamformed data to an external PC to accomplish off-line software processing.

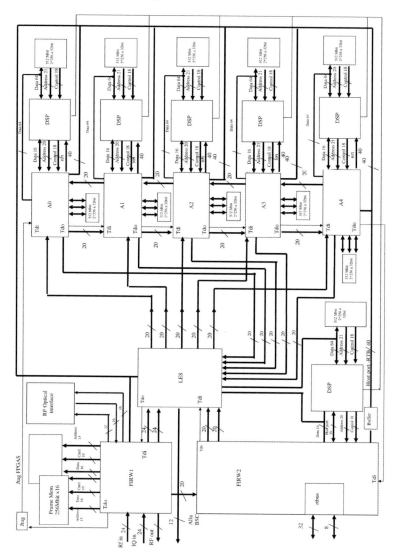

Fig. 13.5. DRP (digital radio-frequency processing) board: block diagram

The first constraint was matched by implementing a PCB based in fully programmable and reconfigurable hardware components. With reference to Figure 13.5, in which a block diagram of the DRP board is reported, the extremely high connectivity of the various blocks as one to every-other can be appreciated. This ensures a high flexibility to re-route the data flow among the various processing components.

The processing blocks are either last generation Virtex4 FPGAs, model sx35, by Xilinx (i.e., FIRW1, FIRW2, LES, A0, A1, A2, A3 and A4) or powerful TMS 6414 DSPs, 600 MHz clock, by Texas Instruments (i.e., DSPs elements).

Connected to FIRW1 there are two 2 M * 32 bit * 4 banks SDRAMs which can be used in ping-pong to manage concurrent memory accesses: in this case, one memory is used to memorize/write the data input stream, whilst the other is used to read previously stored data. Once the data read from this last memory is terminated, the two memories are swapped and their use is changed. The purpose of these two memories is to make available the possibility to store up to (roughly) 8 RF frames to accomplish time correlation processes.

The five Ai-DSP blocks are powerful processing units able to carry out parallel complex processing.

FIRW2, together with the connected DSP, is devoted to implement the interface with the system's bus (i.e., RT bus) and to serve as a master for the board programming.

The second constraint was satisfied by foreseeing the presence of an optical Gigabit Ethernet link (1.8 Gbit/s transfer rate) between the US unit and an external PC used both to control the acquisition and to store the data. The use of such a large bandwidth channel is necessary to transfer the very high-rate RF data flow of the US unit.

The data from the MyLab 70XVG are picked-up just after the receiving beamforming process without any kind of other processing. Actually, the data are picked-up after an RF filtering step which is used to cut-off the residual of DC coming from the A/D conversion process and to limit the bandwidth of the RF data to the real bandwidth of the transducer in use (i.e., a step necessary to reduce the integration of off-bandwidth thermal noise). In this way, the RF filtering step provides a clean RF data flow without blurring it in any way.

The Gigabit Ethernet is a very fast serial link. So, the parallel data from the US data bus have to be serialized. We used an ad hoc 16-to-1 parallel-to-serial converter.

To allow a real-time storing of acquired data, the external PC has to be equipped with a PCI 64 bit bus. This solution is required to have a sufficiently large transmission bandwidth in the "virtual" link between the US and the hard-disk used to store the RF data. Of course, the operative system running on the PC has to be able to manage such a PCI standard. Our choice was to use Linux, being the 64 bit version of the one already widely used, and hence, quite stable and safe from bugs. Particularly, the present hardware/firmware version of the RF acquisition tool runs under Fedora Core 3 64 bits.

The limit of the amount of data which can be acquired is the free space available on the hard-disk.

13.3 Conclusions

In this chapter, a new powerful and flexible US architecture specifically designed to take into account the needs of UCM processing has been described in detail. Particularly, great effort has been devoted to design a very sensitive, fully-linear and high-dynamic front-end, able to process the very small signals reflected by UCM bubbles and to analyze their harmonic contents.

Complex transmission sequences can be set up to deal with multi-firing acquisition techniques (e.g., pulse-inversion [1] and contrast pulse sequencing [1]) and arbitrary transmission waveforms can be treated (i.e., chirp coding [2–5]).

The receiving chain has been designed in order to guarantee the highest possible processing flexibility and reconfiguration capabilities.

Particular care has also been devoted to preserving the largest receiving bandwidth to suitably fit with the possible range of frequencies concerned with UCM scatterings (in practice, from about 500 kHz up to the limit of 20 MHz).

An extended possibility to store both RF and baseband shifted data is also allowed.

Lastly, but even more important, a dedicated board for real-time implement complex RF data processing has been designed and a serial optical link based on Gigabit Ethernet technology has been foreseen to transfer and record acquired data from the scanner, just after the beamforming, to a disk device of an external commercial PC for further off-line processing and/or analysis.

Acknowledgements
This project was funded by the EC in the FP6 programme, priority 2 IST & 3 NMP, under the project no. NMP4-CT-2005-016382, TAMIRUT.

References

1. Hope Simpson D, Chin CT and Burns PN (1999) Pulse Inversion Doppler: A new method for detecting nonlinear echoes from microbubble contrast agents. IEEE Trans. on UFFC 46(2)
2. Misaridis T and Jensen JA (2005) Use of Modulated Excitations Signals in Medical Ultrasound. Part I: Basic Concepts and Expected Benefits. IEEE Trans on UFFC 52(2)
3. Misaridis T and Jensen JA (2005) Use of Modulated Excitations Signals in Medical Ultrasound. Part II: Design and Performance for Medical Imaging Applications. IEEE Trans on UFFC 52(2)
4. Misaridis T and Jensen JA (2005) Use of Modulated Excitations Signals in Medical Ultrasound. Part III: High Frame Rate Imaging. IEEE Trans on UFFC 52(2)
5. Phillips PJ (2001) Contrast pulse sequences (CPS): imaging nonlinear microbubbles. IEEE Ultrasonics Symposium 2:1739–1745
6. Szabo TL (2004) Diagnostic Ultrasound Imaging: Inside Out. Elsevier Academic Press, Amsterdam

Index

End of printing: November 2009